Accelerated Life Testing of One-shot Devices

Accelerated Life Testing of One-shot Devices

Data Collection And Analysis

Narayanaswamy Balakrishnan
McMaster University
Hamilton, Canada

Man Ho Ling
The Education University of Hong Kong
Tai Po, Hong Kong SAR, China

Hon Yiu So
University of Waterloo
Waterloo, Canada

WILEY

This first edition first published 2021

Registered Office
John Wiley & Sons, Inc., 111 River Street, Hoboken, NJ 07030, USA

Editorial Office
111 River Street, Hoboken, NJ 07030, USA

For details of our global editorial offices, customer services, and more information about Wiley products visit us at www.wiley.com.

Wiley also publishes its books in a variety of electronic formats and by print-on-demand. Some content that appears in standard print versions of this book may not be available in other formats.

Library of Congress Cataloging-in-Publication Data

Names: Balakrishnan, Narayanaswamy., 1956- author. | Ling, Man Ho, author. | So, Hon Yiu, author.
Title: Accelerated life testing of one-shot devices : data collection and analysis / Narayanaswamy Balakrishnan, McMaster University, Hamilton, Canada, Man Ho Ling, The Education University of Hong, Kong, New Territories, Hong Kong, Hon Yiu So, University of Waterloo, Waterloo, Canada.
Description: First edition. | Hoboken, NJ, USA : Wiley, 2021. | Includes bibliographical references and index.
Identifiers: LCCN 2020035725 (print) | LCCN 2020035726 (ebook) | ISBN 9781119664000 (cloth) | ISBN 9781119664017 (adobe pdf) | ISBN 9781119663942 (epub)
Subjects: LCSH: Accelerated life testing. | Failure analysis (Engineering)
Classification: LCC TA169.3 .B35 2021 (print) | LCC TA169.3 (ebook) | DDC 620/.00452–dc23
LC record available at https://lccn.loc.gov/2020035725
LC ebook record available at https://lccn.loc.gov/2020035726

Cover Design: Wiley
Cover Image: © Piergiov/Getty Images

FSC www.fsc.org MIX Paper from responsible sources FSC® C013604

With great love and affection, we dedicate this book to

Sarah and Julia Balakrishnan, and Colleen Cutler	*NB*
Grace Chu, Sophia Ling, and Sheldon Ling	*MHL*
Tian Feng and Victoria So	*HYS*

Contents

Preface

Lifetime information obtained from one-shot devices is very limited as the entire data are either left- or right-censored. For this reason, the analysis of one-shot device testing data poses a special challenge. This book provides several statistical inferential methods for analyzing one-shot device lifetime data obtained from accelerated life-tests and also develops optimal designs for two mainstream accelerated life-tests – constant-stress and step-stress accelerated life-tests – that are commonly used in reliability practice. The discussions provided in the book would enable reliability practitioners to better design their experiments for data collection from efficient accelerated life-tests when there are budget constraints in place. This is important from estimation and prediction point of view as such optimal designs would result in as accurate an inference as possible under the constraints imposed on the reliability experiment. Moreover, R codes are presented within each chapter so that users can try out performing their own analysis on one-shot device testing data.

In addition, the inferential methods and the procedures for planning accelerated life-tests discussed in this book are not only limited to one-shot devices alone but also can be extended naturally to accelerated life-tests with periodic inspections (interval-censoring) and those with continuous monitoring and censoring (right-censoring). The book finally concludes by highlighting some important issues and problems that are worth considering for further research. This may be especially useful for research scholars and new researchers interested in taking on this interesting and challenging area of research in reliability theory and practice.

It is possible that some pertinent results or references got omitted in this book, and we assure you that it is due to inadvertency on our part and not due to scientific antipathy. We will appreciate greatly if the readers inform us of any corrections/omissions, or any comments pertinent to any of the discussions in the book!

Our sincere thanks go to the entire Wiley team, Ms. Mindy Okura-Marszycki, Ms. Kathleen Santoloci, and Mr. Brett Kurzman, for taking great interest in this project from day one, for all their help and encouragement during the whole course, and for their fine assistance during the final production stage of the book. Our thanks also go to our research collaborators and graduate students for their incisive comments and queries, which always benefited us greatly and helped clarify some of our own ideas! We express our sincere appreciation to Ms. Elena Maria Castilla Gonzalez, a doctoral student of Professor Leandro Pardo in the Department of Statistics and Operations Research at Complutense University of Madrid, Spain, for her careful reading of Chapter 5 and also for sharing with us some R codes that she had developed concerning robust inferential methods for one-shot device test analyses. Last but not least, our special thanks go to our families for their patience and understanding, and for providing constant support and encouragement during our work on this book!

Finally, the first author (NB) wishes to state to his older daughter, Ms. Sarah Balakrishnan, that though she lost out on getting his Volvo car due to a major car accident, she should be heartened by the fact that the accident resulted in the germination of his interest and ideas on one-shot devices (airbags), and ultimately this book solely dedicated to the topic!

July, 2020

Narayanaswamy Balakrishnan
Man Ho Ling
Hon Yiu So

About the Companion Website

This book is accompanied by a companion website:

www.wiley.com/go/Balakrishnan/Accelerated_Life_Testing

The Student companion site will contain the codes and case studies.

1

One-Shot Device Testing Data

1.1 Brief Overview

One-shot device testing data analyses have recently received great attention in reliability studies. The aim of this chapter is to provide an overview on one-shot device testing data collected from accelerated life-tests (ALTs). Section 1.2 surveys typical examples of one-shot devices and associated tests in practical situations. Section 1.3 describes several popular ALTs, while Section 1.4 provides some examples of one-shot device testing data that are typically encountered in reliability and survival studies. Finally, Section 1.5 details some recent developments on one-shot device testing data analyses and associated issues of interest.

1.2 One-Shot Devices

Valis et al. (2008) defined one-shot devices as units that are accompanied by an irreversible chemical reaction or physical destruction and could no longer function properly after its use. Many military weapons are examples of one-shot devices. For instance, the mission of an automatic weapon gets completed successfully only if it could fire all the rounds placed in a magazine or in ammunition feed belt without any external intervention. Such devices will usually get destroyed during usual operating conditions and can therefore perform their intended function only once.

Shaked and Singpurwalla (1990) discussed the submarine pressure hull damage problem from a Bayesian perspective and assessed the effect of various strengths of underwater shock waves caused by either a nuclear device or a chemical device on the probability of damage to a submarine pressure hull. A record is made of whether a copy of a diminutive model of a submarine pressure hull is damaged or not, and a specific strength of the shock wave on the model. Fan et al. (2009)

Accelerated Life Testing of One-shot Devices: Data Collection and Analysis, First Edition.
Narayanaswamy Balakrishnan, Man Ho Ling, and Hon Yiu So.

Companion Website: www.wiley.com/go/Balakrishnan/Accelerated_Life_Testing

considered electro-explosive devices in military applications, which induct a current to excite inner powder and make them explode. Naturally, we cannot adjudge the functioning condition of the electro-explosive device from its exterior, but can only observe it by detonating it directly. After a successful detonation, the device cannot be used anymore; if the detonation becomes a failure, we will also not know when exactly it failed. Nelson (2003) described a study of crack initiation for turbine wheels. Each of the 432 wheels was inspected once to determine whether it had started to crack or not. Newby (2008) provided some other examples of one-shot devices, such as fire extinguishers or munitions. A full test would require the use of the considered devices and, therefore, their subsequent destruction. The test carried out would show whether a device is still in a satisfactory state, or has failed by that inspection time.

One-shot device testing data also arise in destructive inspection procedures, wherein each device is allowed for only a single inspection because the test itself results in its destruction. Morris (1987) presented a study of 52 Li/SO_2 storage batteries under destructive discharge. Each battery was tested at one of three inspection times and then classified as acceptable or unacceptable according to a critical capacity value.

Ideally, reliability data would contain actual failure times of all devices placed on test (assuming, of course, the experimenter could wait until all devices fail), so that the observed failure times can reveal the failure pattern over time, and we could then estimate the reliability of the device reasonably. But, in practice, many life-tests would get terminated before all the units fail. Such an early stoppage of the life-test by the experimenter may be due to cost or time constraints or both. This would result in what is called as "right-censored data" because the exact failure times of the unfailed devices are unknown, but all we know is that the failure times of those devices are larger than the termination time. Considerable literature exists on statistical inference for reliability data under right-censoring; for example one may refer to the books by Cohen (1991), Balakrishnan and Cohen (1991), and Nelson (2003).

Moreover, when nondestructive and periodic inspections are carried on devices, their exact failure times will not be observed, but the intervals wherein the failures occurred will only be available. If a failure is observed by the first inspection, then it is known that the failure time of the device is less than the first inspection time, resulting in "left-censoring." Similarly, if a failure is observed between two consecutive inspection times, then it is known that the failure time is between these two corresponding inspection times, resulting in "interval-censoring." Finally, the failure times of all surviving units at the final inspection time are right-censored as their exact failure times will not be observed. Exact failure times can only be observed from a life-testing experiment with continuous surveillance. The periodic inspection process with nondestructive evaluation would actually provide a reasonable approximation to failure times of devices under test, especially when the inspection time intervals are short, even though the precision of inference will be less in this case.

It is useful to note that in all the preceding examples of one-shot devices, we will not observe the actual lifetimes of the devices. Instead, we would only observe either a success or a failure at the inspection times, and so only the corresponding binary data would be observed, consequently resulting in less precise inference. In this manner, one-shot device testing data differ from typical data obtained by measuring lifetimes in standard life-tests and, therefore, poses a unique challenge in the development of reliability analysis, due to the lack of lifetime information being collected from reliability experiments on such one-shot devices. If successful tests occur, it implies that the lifetimes are beyond the inspection times, leading to right-censoring. On the other hand, the lifetimes are before the inspection times, leading to left-censoring, if tests result in failures. Consequently, all lifetimes are either left- or right-censored. In such a setting of the lifetime data, Hwang and Ke (1993) developed an iterative procedure to improve the precision of the maximum likelihood estimates for the three-parameter Weibull distribution and to evaluate the storage life and reliability of one-shot devices. Some more examples of one-shot devices in the literature include missiles, rockets, and vehicle airbags; see, for example, Bain and Engelhardt (1991), Guo et al. (2010), and Yun et al. (2014).

1.3 Accelerated Life-Tests

As one-shot devices (such as ammunition or automobile airbags) are usually kept for a long time in storage and required to perform its function only once, the reliability required from such devices during their normal operating conditions would naturally be high. So, it would be highly unlikely to observe many failures on tests under normal operating conditions within a short period of time. This renders the estimation of reliability of devices to be a challenging problem from a statistical point of view. In this regard, ALTs could be utilized to mitigate this problem. In ALTs, devices are subject to higher-than-normal stress levels to induce early failures. In this process, more failures could likely be obtained within a limited test time. As the primary goal of the analysis is to estimate the reliability of devices under normal operating conditions, ALT models would then typically extrapolate (from the data obtained at elevated stress levels) to estimate the reliability under normal operating conditions. ALTs are known to be efficient in capturing valuable lifetime information, especially when there is a need to shorten the life-testing experiment. For this reason, ALTs have become popular and are commonly adopted in many reliability experiments in practice. One may refer to the detailed reviews presented by Nelson (1980), Cramer and Kamps (2001), Pham (2006), and Meeker and Escobar (2014), and the excellent booklength account provided by Nelson (2009).

Constant-stress accelerated life-tests (CSALTs) and step-stress accelerated life-tests (SSALTs) are two popular ALT plans that have received great attention in the literature. Under a CSALT, each device gets tested at only one prespecified stress level. To mention a few recent works, for example, Wang et al. (2014) considered CSALTs with progressively Type-II right censored samples under Weibull lifetime distribution; for pertinent details on progressive censoring, see Balakrishnan (2007) and Balakrishnan and Cramer (2014). Wang (2017) discussed CSALTs with progressive Type-II censoring under a lower truncated distribution. Lin et al. (2019) studied CSALTs terminated by a hybrid Type-I censoring scheme under general log-location-scale lifetime distributions. SSALTs are an alternative to apply stress to devices in a way that stress levels will increase at prespecified times step-by-step. For SSALTs, there are three fundamental models for the effect of increased stress levels on the lifetime distribution of a device: The tampered random variable model proposed by DeGroot and Goel (1979), the cumulative exposure model of Sedyakin (1966) and Nelson (1980); see also (Nikulin and Tahir, 2013), and the tampered failure rate model proposed by Bhattacharyya and Soejoeti (1989). All these models of SSALTs have been discussed extensively by many authors. Gouno (2001) analyzed data collected from SSALTs and presented an optimal design for SSALTs; see also Gouno (2007). Zhao and Elsayed (2005) analyzed data on the light intensity of light emitting diodes collected from SSALTs with four stress levels under Weibull and log-normal distributions. For the case of exponential lifetime distribution, by considering a simple SSALT under Type-II censoring, Balakrishnan et al. (2007) developed exact likelihood inferential methods for the model parameters; see also Balakrishnan (2008) for details, while Xiong et al. (2006) considered the situation when the stress changes from a low-level stress to a high-level stress at a random time.

1.4 Examples in Reliability and Survival Studies

1.4.1 Electro-Explosive Devices Data

Fan et al. (2009) considered data, presented in Table 1.1, on 90 electro-explosive devices under various levels of temperature at different inspection times. Ten devices under test at each condition were inspected to see whether there were any

Table 1.1 Failure records on electro-explosive devices under CSALTs with temperature (K).

Test group	Inspection time	Temperature	Number of samples	Number of failures
1	10	308	10	3
2	10	318	10	1
3	10	328	10	6
4	20	308	10	3
5	20	318	10	5
6	20	328	10	7
7	30	308	10	7
8	30	318	10	7
9	30	328	10	9

Source: Fan et al. (2009).

failures or not at each inspection time for each temperature setting. These data were then used to estimate the reliability of electro-explosive devices at different mission times under the normal operating temperature.

1.4.2 Glass Capacitors Data

Zelen (1959) presented data from a life-test of glass capacitors at four higher-than-usual levels of temperature and two levels of voltage. At each of the eight combinations of temperature and voltage, eight items were tested. We adopt these data to form one-shot device testing data by taking the inspection times (hours) as $\tau \in \{300, 350, 400, 450\}$, which are summarized in Table 1.2. These data were then used to estimate the mean lifetime of glass capacitors for 250 V and 443 K temperature.

1.4.3 Solder Joints Data

Lau et al. (1988) considered data on 90 solder joints under three types of printed circuit boards (PCBs) at different temperatures. The lifetime was measured as the number of cycles until the solder joint failed, while the failure of a solder joint is defined as a 10% increase in measured resistance. A simplified dataset is derived from the original one and presented in Table 1.3, where two stress factors considered are temperature and a dichotomous variable indicating if the PCB type is "copper-nickel-tin" or not.

Table 1.2 Failure records on glass capacitors under CSALTs with two stress factors: temperature (K) and voltage (V).

Test group	Inspection time	Temperature	Voltage	Number of samples	Number of failures
1	450	443	200	8	1
2	400	453	200	8	0
3	350	443	250	8	0
4	300	453	250	8	1
5	450	443	300	8	3
6	400	453	300	8	4
7	350	443	350	8	3
8	300	453	350	8	2

Source: Zelen (1959).

Table 1.3 Failure records on solder joints under CSALTs with temperature (K) and a dichotomous variable indicating if the PCB type is "copper-nickel-tin (CNT)" or not.

Test group	Inspection time	Temperature	CNT	Number of samples	Number of failures
1	300	293	Yes	10	4
2	300	333	Yes	10	4
3	100	373	Yes	10	6
4	1300	293	No	20	10
5	800	333	No	20	3
6	200	373	No	20	4

Source: Lau et al. (1988).

1.4.4 Grease-Based Magnetorheological Fluids Data

Zheng et al. (2018) studied grease-based magnetorheological fluids under SSALTs with four levels of temperature and observed whether their viscosities or shear stresses decreased by more than 10% after tests. Twenty samples of grease-based magnetorheological fluids were subject to higher-than-normal operating temperature. Then, each sample was inspected only once and only whether it had failed or not at the inspection time was observed, and not the actual failure time. The data collected in this manner, presented in Table 1.4, were then used to estimate the mean lifetime of grease-based magnetorheological fluids under the normal operating temperature.

1.4.5 Mice Tumor Toxicological Data

It is important to point out that one-shot device testing data arise from diverse fields beyond reliability engineering, such as in mice tumor studies from tumorigenicity experiments; see Kodell and Nelson (1980). In such a study, each mouse received a particular dosage of benzidine dihydrochloride in its drinking water and was later sacrificed to detect whether some tumors had developed by then or not. Tumor presence can be detected only at the time of mouse's sacrifice or natural death. These data are summarized in Table 1.5. The data collected in this form were then used to measure the impact of the chemical dosage on the risk of tumor development.

1.4.6 ED01 Experiment Data

Lindsey and Ryan (1993) described experimental results conducted by National Center for Toxicological Research in 1974. 3355 out of 24 000 female mice were randomized to a control group or groups that were injected with a high dose (150 ppm) of a known carcinogen, called 2-AAF, to different parts of the bodies. The inspection times on the mice were 12, 18, and 33 months and the outcomes of mice were death without tumor (DNT) and death with tumor (DWT), and sacrificed without tumor (SNT) and sacrificed with tumor (SWT). Balakrishnan et al. (2016a), in their analysis, ignored the information about parts of mouse bodies where the drugs were injected and combined SNT and SWT into one sacrificed group, and denoted the cause of DNT as natural death and the cause of DWT as death due to cancer. These data are summarized in Table 1.6. They then estimated the chance of death without tumor.

1.4.7 Serial Sacrifice Data

Ling et al. (2020) were primarily concerned with the data (Berlin et al., 1979), presented in Table 1.7, on the presence or absence of two disease categories – (a)

Table 1.4 Failure records on grease-based magnetorheological fluids under SSALTs with temperature (K).

Stage	Inspection time (h)	Temperature	Number of samples	Number of failures
1	864	333	5	1
2	1512	339	5	1
3	1944	345	5	2
4	2160	351	5	2

Source: Zheng et al. (2018).

Table 1.5 The number of mice sacrificed, with tumor from tumorigenecity experiments data.

Test group	Inspection time (mo)	Sex	Dosage (ppm)	Number of mice sacrificed	Number of mice with tumor
1	9.33	F	60	72	1
2	14.00	F	60	48	3
3	18.67	F	60	36	18
4	9.33	F	120	48	0
5	14.00	F	120	47	14
6	18.67	F	120	26	25
7	9.33	F	200	47	4
8	14.00	F	200	45	38
9	9.33	F	400	24	16
10	14.00	F	400	10	9
11	9.33	M	120	48	0
12	14.00	M	120	44	7
13	18.67	M	120	42	11
14	9.33	M	200	47	3
15	14.00	M	200	32	5
16	18.67	M	200	19	8
17	9.33	M	400	24	0
18	14.00	M	400	22	11
19	18.67	M	400	15	11

Source: Kodell and Nelson (1980).

thymic lymphoma and/or glomerulosclerosis and (b) all other diseases – for an irradiated group of 343 female mice given γ-radiation and a control group of 361 radiation-free female mice to study the onset time and the rate of development of radiation-induced disease. All of the mice in both groups were sacrificed at various times, with the presence of a disease indicating that the disease onset occurred before sacrifice, while the absence of a disease indicating that the disease onset would occur after sacrifice.

Table 1.6 The number of mice sacrificed, died without tumor, and died with tumor from the ED01 experiment data.

Test group	Inspection time (mo)	High dose of 2-AAF	Number of mice		
			Sacrificed	Died without tumor	Died with tumor
1	12	No	115	22	8
2	12	Yes	110	49	16
3	18	No	780	42	8
4	18	Yes	540	54	26
5	33	No	675	200	85
6	33	Yes	510	64	51

Source: Lindsey and Ryan (1993).

Table 1.7 Serial sacrifice data on the presence or absence of two disease categories: (a) thymic lymphoma and/or glomerulosclerosis and (b) all other diseases.

Test group	Sacrifice time (d)	γ-radiation	Number of mice			
			Healthy	With (a) only	With (b) only	With (a) and (b)
1	100	No	58	13	0	1
2	200	No	40	23	1	1
3	300	No	18	41	1	3
4	400	No	8	25	1	6
5	500	No	1	21	1	16
6	600	No	1	11	0	21
7	700	No	0	9	1	39
8	100	Yes	54	12	1	0
9	200	Yes	36	24	3	5
10	300	Yes	13	35	1	17
11	400	Yes	0	13	2	28
12	500	Yes	0	3	1	35
13	600	Yes	0	0	1	30
14	700	Yes	0	0	1	28

Source: Berlin et al. (1979).

1.5 Recent Developments in One-Shot Device Testing Analysis

We now provide a brief review of some recent developments on one-shot device testing data analyses under ALTs. For CSALTs, Fan et al. (2009) compared three different prior distributions in the Bayesian approach for making predictions on the reliability at a mission time and the mean lifetime of electro-explosive devices under normal operating conditions. In a series of papers, Balakrishnan and Ling (2012a,b, 2013, 2014a) developed expectation-maximization (EM) algorithms for the maximum likelihood estimation of model parameters based on one-shot device testing data under exponential, Weibull and gamma lifetime distributions. In addition to parameter estimation, different methods of confidence intervals for the mean lifetime and the reliability at a mission time under normal operating conditions have also been discussed by these authors. The maximum likelihood estimation as well as associated tests of hypotheses, though are most efficient when the assumed model is indeed the true model, are known to be non-robust when the assumed model is violated, for example, by the presence of some outlying values in the data. With this in mind, weighted minimum density power divergence estimators have recently been developed for one-shot device testing data under exponential, Weibull, and gamma distributions (Balakrishnan et al., 2019a,b, 2020a,b).

Some other important inferential aspects, apart from the works on the estimation of model parameters and hypothesis tests described above, have also been addressed by a number of authors. Balakrishnan and Ling (2014b), for instance, have developed a procedure to obtain CSALT plans when there are budget constraints for testing one-shot devices. Ling and Balakrishnan (2017) also studied model mis-specification effects on one-shot device testing data analyses between Weibull and gamma distributions, while Balakrishnan and Chimitova (2017) conducted comprehensive simulation studies to compare the performance of several goodness-of-fit tests for one-shot device testing data.

In the framework of competing risks analysis, Balakrishnan et al. (2015, 2016a,b) discussed the analysis of one-shot device testing data when the devices contain multiple components and hence having multiple failure modes. Another extension that has been provided for one-shot device testing is by Ling et al. (2016) who have developed proportional hazards models for analyzing such data. Optimal

SSALT plans for one-shot device testing experiment with lifetimes following exponential and Weibull distributions have been discussed by Ling (2019) and Ling and Hu (2020).

Pan and Chu (2010) have investigated two- and three-stage inspection schemes for assessing one-shot devices in series systems of components having Weibull lifetime distributions. Finally, Cheng and Elsayed (2016–2018) have examined several approaches to measure the reliability of one-shot devices with mixture of units under various scenarios and have presented reliability metrics of systems with mixtures of nonhomogeneous one-shot units subject to thermal cyclic stresses and further optimal operational use of such systems.

In the chapters that follow, we shall elaborate on all these developments and also highlight their applications.

2

Likelihood Inference

2.1 Brief Overview

Likelihood inference is one of the classical methods for the estimation of model parameters, and it relies on the maximization of likelihood function of an assumed model based on the observed data. In this chapter, we detail the procedure of finding the maximum likelihood estimates (MLEs) of model parameters, mean lifetime, and reliability under normal operating conditions on the basis of one-shot device testing data under constant-stress accelerated life-tests (CSALTs) for different lifetime distributions. Several associated interval estimation methods are also discussed for all lifetime parameters of interest. The discussions provided here are primarily from the works carried out by Balakrishnan and Ling (2012a,b, 2013, 2014a). Then, the results of a simulation study evaluating the performance of the developed point and interval estimation methods are presented. Finally, two data sets presented in the last chapter are used to illustrate all the inferential results discussed here.

2.2 Under CSALTs and Different Lifetime Distributions

Suppose CSALTs are performed on I groups of one-shot devices. Suppose for $i = 1, 2, \dots, I$, in the ith test group, K_i one-shot devices are subject to J types of accelerating factors (such as temperature, humidity, voltage, and pressure) at stress levels $\mathbf{x}_i = (x_{i1}, x_{i2}, \dots, x_{iJ})$ and get inspected at time τ_i for their condition. The corresponding numbers of failures, n_i, are then recorded. The data thus observed can be summarized as in Table 2.1.

For notational convenience, we use $\mathbf{z} = \{\tau_i, K_i, n_i, \mathbf{x}_i, i = 1, 2, \dots, I\}$ to denote the observed data, and $\boldsymbol{\theta}$ for the model parameter. Suppose the lifetime in the ith test group has a distribution with cumulative distribution function (cdf) $F(t; \boldsymbol{\Psi}_i)$, with

Accelerated Life Testing of One-shot Devices: Data Collection and Analysis, First Edition.
Narayanaswamy Balakrishnan, Man Ho Ling, and Hon Yiu So.

Companion Website: www.wiley.com/go/Balakrishnan/Accelerated_Life_Testing

Table 2.1 One-shot device testing data under CSALTs with multiple accelerating factors, various stress levels, and different inspection times.

Test group	Stress levels	Inspection time	Number of tested devices	Number of failures
1	$(x_{11}, x_{12}, \ldots, x_{1J})$	τ_1	K_1	n_1
2	$(x_{21}, x_{22}, \ldots, x_{2J})$	τ_2	K_2	n_2
$\vdots$	$\vdots$	$\vdots$	$\vdots$	$\vdots$
I	$(x_{I1}, x_{I2}, \ldots, x_{IJ})$	τ_I	K_I	n_I

$\boldsymbol{\Psi}_i = \boldsymbol{\Psi}(\mathbf{x}_i; \boldsymbol{\theta})$. The observed likelihood function is then given by

$$L(\boldsymbol{\theta}; \mathbf{z}) = C \prod_{i=1}^{I} [F(\tau_i; \boldsymbol{\Psi}_i)]^{n_i} [R(\tau_i; \boldsymbol{\Psi}_i)]^{K_i - n_i}, \tag{2.1}$$

where $R(\tau_i; \boldsymbol{\Psi}_i) = 1 - F(\tau_i; \boldsymbol{\Psi}_i)$ is the reliability at inspection time τ_i and C is the normalizing constant. We then make use of the data to determine some lifetime characteristics of devices, from the likelihood function in (2.1), under normal operating conditions $\mathbf{x}_0 = (x_{01}, x_{02}, \ldots, x_{0J})$, such as the model parameter, $\boldsymbol{\theta}$, mean lifetime, μ, and reliability at mission time t, $R(t)$.

2.3 EM-Algorithm

Likelihood inference on one-shot device testing data has been discussed extensively for many prominent lifetime distributions; see Balakrishnan and Ling (2012a,b, 2013,2014a). In this section, we describe the expectation-maximization (EM) framework for determining the MLEs of all parameters of interest.

In one-shot device testing, as mentioned earlier, no actual lifetimes are observed, and, as such, all observed data are censored. EM algorithm is known to be a convenient and efficient method for estimating model parameters in the presence of censoring; see, for example, McLachlan and Krishnan (2008) for all pertinent details concerning this method, its variations, and applications. This method has now become a standard method of model-fitting in the presence of missing data. The EM algorithm involves two steps in each iteration of the numerical method of maximizing the likelihood function such as the one presented in (2.1): expectation-step (E-step) in which the missing data are approximated by their expected values, and maximization-step (M-step) in which the likelihood function, with imputed values replacing the missing data, gets maximized.

Suppose the lifetimes of devices follow a distribution with probability density function (pdf) $f(t;\boldsymbol{\theta})$. We first note that the log-likelihood function, based on the complete data, is

$$\ell_c(\boldsymbol{\theta}) = \sum_{i=1}^{I}\sum_{k=1}^{K_i} \ln(f(t_{ik};\boldsymbol{\theta})). \tag{2.2}$$

In the mth step of the iterative process, the objective then is to update the estimate of parameter $\boldsymbol{\theta}$ by the value that maximizes the function

$$Q(\boldsymbol{\theta},\boldsymbol{\theta}^{(m)}) = E_{\boldsymbol{\theta}^{(m)}}[\ell_c(\boldsymbol{\theta})|\mathbf{z}], \tag{2.3}$$

based on the current estimate $\boldsymbol{\theta}^{(m)}$. The current estimate $\boldsymbol{\theta}^{(m)}$ is used in obtaining the updated estimate $\boldsymbol{\theta}^{(m+1)}$ in the M-step, and then the conditional expectation $Q(\boldsymbol{\theta},\boldsymbol{\theta}^{(m+1)})$ in (2.3) is obtained based on $\boldsymbol{\theta}^{(m+1)}$ in the E-step. These two steps are then repeated until convergence is achieved to a desired level of accuracy. It is evident that we actually solve the incomplete data problem of maximizing the likelihood function in (2.1) by first approximating the missing data and then using the approximated values to find the estimate of the parameter vector as the solution for the complete data problem.

However, in the maximization step, a closed-form solution may not always be found, and in such a case, a one-step Newton–Raphson method can be employed for this purpose. This would require the second-order derivatives of the log-likelihood function with respect to the model parameters. In the present case, let us introduce

$$\mathbf{I}^{(m)} = \left[\frac{\partial Q(\boldsymbol{\theta},\boldsymbol{\theta}^{(m)})}{\partial\boldsymbol{\theta}}\right]_{\boldsymbol{\theta}=\boldsymbol{\theta}^{(m)}} \tag{2.4}$$

and

$$\mathbf{J}^{(m)} = -\left[\frac{\partial^2 Q(\boldsymbol{\theta},\boldsymbol{\theta}^{(m)})}{\partial\boldsymbol{\theta}\,\partial\boldsymbol{\theta}'}\right]_{\boldsymbol{\theta}=\boldsymbol{\theta}^{(m)}}. \tag{2.5}$$

Then, the updated estimate obtained through the one-step Newton–Raphson method is given by

$$\boldsymbol{\theta}^{(m+1)} = (\mathbf{J}^{(m)})^{-1}\mathbf{I}^{(m)} + \boldsymbol{\theta}^{(m)}. \tag{2.6}$$

In addition to estimates of model parameters, our interest may also be on some lifetime characteristics, such as mean lifetime and reliability under normal operating conditions. As these quantities are functions of model parameters, their estimates can be readily found by plugging in the MLEs of model parameters into the respective functions, and then the delta method can be employed to determine the corresponding standard errors as well; see, for example, Casella and Berger (2002).

2.3.1 Exponential Distribution

Let us first start with the exponential distribution, which is one of the most popular lifetime distributions; see, for example, Johnson et al. (1994) and Balakrishnan and Basu (1995) for developments on various methods and applications of the exponential distribution. In this case, Fan et al. (2009) and Balakrishnan and Ling (2012a,b) studied inferential methods for one-shot device testing data under exponential distribution with single and multiple accelerating factors. Let us use E to denote the exponential distribution in what follows. In this case, in the ith test group, the cdf and pdf of the exponential distribution, with rate parameter $\lambda_i > 0$, are given by

$$F_E(t; \lambda_i) = 1 - \exp(-\lambda_i t), \quad t > 0, \tag{2.7}$$

and

$$f_E(t; \lambda_i) = \frac{\partial F_E(t; \lambda_i)}{\partial t} = \lambda_i \exp(-\lambda_i t), \quad t > 0, \tag{2.8}$$

respectively. Its mean lifetime and reliability at mission time t under normal operating conditions $\mathbf{x}_0 = (x_{01}, x_{02}, \ldots, x_{0J})$ are then given by

$$\mu_E = \int_0^\infty t f_E(t; \lambda_0) dt = \frac{1}{\lambda_0}$$

and

$$R_E(t) = 1 - F_E(t; \lambda_0) = \exp(-\lambda_0 t),$$

respectively. As it is necessary to extrapolate the data observed from the elevated stress levels $\mathbf{x}_i = (x_{i1}, x_{i2}, \ldots, x_{iJ})$, to lifetime characteristics of devices under normal operating conditions $\mathbf{x}_0 = (x_{01}, x_{02}, \ldots, x_{0J})$, the rate parameter λ_i in (2.7) gets related to the stress levels in a log-linear link function of the form, with $x_{i0} = 1$,

$$\ln(\lambda_i) = \sum_{j=0}^{J} e_j x_{ij}, \tag{2.9}$$

where $\boldsymbol{\theta}_E = (e_0, e_1, \ldots, e_J)$ now becomes the new model parameter.

It is worth mentioning that many well-known stress-rate models, such as Arrhenius, power law, inverse power law, and Eyring models, are all special cases of the log-linear link function in (2.9), with suitable transformations on the stress variables; see, for example, Wang and Kececioglu (2000).

Under the EM framework, the log-likelihood function based on the complete data, obtained readily from (2.8), is given by

$$\ell_c = \sum_{i=1}^{I} \sum_{k=1}^{K_i} \ln(f_E(t_{ik}; \lambda_i)) = \sum_{i=1}^{I} \left\{ K_i \sum_{j=0}^{J} e_j x_{ij} - \exp\left(\sum_{j=0}^{J} e_j x_{ij} \right) \sum_{k=1}^{K_i} t_{ik} \right\},$$

and the conditional expectation in (2.3) becomes

$$Q(\boldsymbol{\theta}_E, \boldsymbol{\theta}_E^{(m)}) = \sum_{i=1}^{I} K_i \left\{ \sum_{j=0}^{J} e_j x_{ij} - \exp\left(\sum_{j=0}^{J} e_j x_{ij} \right) A_i^{(m)} \right\}, \tag{2.10}$$

where $A_i^{(m)} = E[T_i | \mathbf{z}, \theta_E^{(m)}]$ can be computed in the E-step, as described below.

The first-order derivatives with respect to the model parameters, required for the maximization of the quantity $Q(\boldsymbol{\theta}_E, \boldsymbol{\theta}_E^{(m)})$ in (2.10), are then given by

$$\frac{\partial Q(\boldsymbol{\theta}_E, \boldsymbol{\theta}_E^{(m)})}{\partial e_j} = \sum_{i=1}^{I} K_i x_{ij} (1 - \lambda_i A_i^{(m)}), \quad j = 0, 1, \ldots, J. \tag{2.11}$$

A closed-form solution is not available for (2.11), however. So, in the M-step of the EM algorithm, we employ the one-step Newton–Raphson method, as mentioned earlier. This requires the second-order derivatives of (2.10) with respect to the model parameters, which are given by

$$\frac{\partial^2 Q(\boldsymbol{\theta}_E, \boldsymbol{\theta}_E^{(m)})}{\partial e_p \, \partial e_q} = -\sum_{i=1}^{I} K_i x_{ip} x_{iq} \lambda_i A_i^{(m)}, \quad p, q = 0, 1, \ldots, J.$$

Next, the conditional expectation required in the E-step is easy to derive as it only involves the failed and unfailed devices. For the failed devices, the proportion is n_i/K_i, and the lifetimes in this case are right-truncated at the inspection time τ_i (as they cannot be more than τ_i). Similarly, for the unfailed devices, the proportion is $1 - n_i/K_i$, and the lifetimes are left-truncated at the inspection time τ_i (as they cannot be less than τ_i). Consequently, we obtain the conditional expectation of T_i, given the observed data and the current estimate $\boldsymbol{\theta}_E^{(m)}$ of $\boldsymbol{\theta}_E$, as follows:

$$\begin{aligned}
A_i^{(m)} &= E[T_i | \mathbf{z}, \boldsymbol{\theta}_E^{(m)}] \\
&= \left(\frac{n_i}{K_i}\right) \frac{\int_0^{\tau_i} t f_E(t; \boldsymbol{\theta}_E^{(m)}) dt}{F_E(\tau_i; \boldsymbol{\theta}_E^{(m)})} + \left(1 - \frac{n_i}{K_i}\right) \frac{\int_{\tau_i}^{\infty} t f_E(t; \boldsymbol{\theta}_E^{(m)}) dt}{1 - F_E(\tau_i; \boldsymbol{\theta}_E^{(m)})} \\
&= \left(\frac{n_i}{K_i}\right) \left(\frac{1}{\lambda_i^{(m)}} - \frac{\tau_i \exp(-\lambda_i^{(m)} \tau_i)}{F_E(\tau_i; \boldsymbol{\theta}_E^{(m)})} \right) + \left(1 - \frac{n_i}{K_i}\right) \left(\tau_i + \frac{1}{\lambda_i^{(m)}} \right) \\
&= \frac{1}{\lambda_i^{(m)}} + \tau_i \left(1 - \frac{\hat{F}(\tau_i)}{F_E(\tau_i; \boldsymbol{\theta}_E^{(m)})} \right),
\end{aligned} \tag{2.12}$$

where $\lambda_i^{(m)} = \exp(\sum_{j=0}^{J} e_j^{(m)} x_{ij})$ and $\hat{F}(\tau_i) = n_i/K_i$ is the empirical probability of failure by inspection time τ_i.

When the MLEs of the model parameters $\hat{\boldsymbol{\theta}}_E = (\hat{e}_0, \hat{e}_1, \ldots, \hat{e}_J)$ are determined, we can readily have the MLEs of mean lifetime and reliability at mission time t under normal operating conditions $\mathbf{x}_0 = (x_{01}, x_{02}, \dots, x_{0J})$, for example, to be

$$\hat{\mu}_E = \frac{1}{\hat{\lambda}_0} = \exp\left(-\sum_{j=0}^{J} \hat{e}_j x_{0j}\right)$$

and

$$\hat{R}_E(t) = \exp(-\hat{\lambda}_0 t) = \exp\left\{-\exp\left(\sum_{j=0}^{J} \hat{e}_j x_{0j}\right) t\right\},$$

respectively.

2.3.2 Gamma Distribution

As the exponential distribution would only be suitable for modeling constant hazard rate, it would be natural to consider the life distribution to be gamma as it possesses increasing, decreasing, and constant hazard rate properties; see, for example, Johnson et al. (1994). Moreover, it includes exponential distribution as a particular case. For this reason, Balakrishnan and Ling (2014a) extended the work to gamma lifetime distribution with multiple accelerating factors with both scale and shape parameters varying over stress factors.

Let us use G to denote gamma distribution in what follows. Specifically, in the ith test group, we assume the lifetimes to follow a gamma distribution with shape parameter $\alpha_i > 0$ and scale parameter $\beta_i > 0$, with cdf and pdf given by

$$F_G(t; \alpha_i, \beta_i) = \gamma\left(\alpha_i, \frac{t}{\beta_i}\right), \quad t > 0, \tag{2.13}$$

and

$$f_G(t; \alpha_i, \beta_i) = \frac{\partial F_G(t; \alpha_i, \beta_i)}{\partial t} = \frac{t^{\alpha_i - 1}}{\Gamma(\alpha_i)\beta_i^{\alpha_i}} \exp\left(-\frac{t}{\beta_i}\right), \quad t > 0, \tag{2.14}$$

respectively, where $\gamma(z, s) = \int_0^s x^{z-1} \exp(-x) dx / \Gamma(z)$ is the lower incomplete gamma ratio and $\Gamma(z) = \int_0^\infty x^{z-1} \exp(-x) dx$ is the complete gamma function. The mean lifetime and reliability at mission time t under normal operating conditions $\mathbf{x}_0 = (x_{01}, x_{02}, \dots, x_{0J})$ are, respectively, given by

$$\mu_G = \int_0^\infty t f_G(t; \alpha_0, \beta_0) dt = \alpha_0 \beta_0$$

and

$$R_G(t) = 1 - F_G(t; \alpha_0, \beta_0) = \Gamma\left(\alpha_0, \frac{t}{\beta_0}\right),$$

where $\Gamma(z, s) = 1 - \gamma(z, s)$ is the upper incomplete gamma ratio. For extrapolating the data observed from the elevated stress levels $\mathbf{x}_i = (x_{i1}, x_{i2}, \dots, x_{iJ})$

to lifetime characteristics of devices under normal operating conditions $\mathbf{x}_0 = (x_{01}, x_{02}, \dots, x_{0J})$, in this case, both parameters α_i and β_i are related to the stress levels in log-linear link functions of the form, with $x_{i0} = 1$,

$$\ln(\alpha_i) = \sum_{j=0}^{J} a_j x_{ij} \quad \text{and} \quad \ln(\beta_i) = \sum_{j=0}^{J} b_j x_{ij}. \tag{2.15}$$

Thus, the new model parameter of the considered model is

$$\boldsymbol{\theta}_G = (a_0, a_1, \dots, a_J, b_0, b_1, \dots, b_J).$$

Here, under the EM framework, the log-likelihood function, based on the complete data, is readily obtained from (2.14) to be

$$\begin{aligned}\ell_c &= \sum_{i=1}^{I}\sum_{k=1}^{K_i} \ln(f_G(t_{ik}; \alpha_i, \beta_i))\\ &= \sum_{i=1}^{I}\left(-K_i \ln(\Gamma(\alpha_i)) - K_i\alpha_i \ln(\beta_i) + (\alpha_i - 1)\sum_{k=1}^{K_i} \ln(t_{ik}) - \frac{1}{\beta_i}\sum_{k=1}^{K_i} t_{ik}\right).\end{aligned}$$

Let $A_i^{(m)} = E[\ln(T_i)|\mathbf{z}, \boldsymbol{\theta}_G^{(m)}]$ and $B_i^{(m)} = E[T_i|\mathbf{z}, \boldsymbol{\theta}_G^{(m)}]$. Then, the conditional expectation in (2.3) becomes

$$Q(\boldsymbol{\theta}_G, \boldsymbol{\theta}_G^{(m)}) = \sum_{i=1}^{I} K_i\left(-\ln(\Gamma(\alpha_i)) - \alpha_i \ln(\beta_i) + (\alpha_i - 1)A_i^{(m)} - \frac{B_i^{(m)}}{\beta_i}\right). \tag{2.16}$$

The first-order derivatives with respect to the model parameters, required for the maximization of the quantity $Q(\boldsymbol{\theta}_G, \boldsymbol{\theta}_G^{(m)})$ in (2.16), are given by

$$\begin{aligned}\frac{\partial Q(\boldsymbol{\theta}_G, \boldsymbol{\theta}_G^{(m)})}{\partial a_j} &= -\sum_{i=1}^{I} K_i x_{ij}\alpha_i(\Psi(\alpha_i) + \ln(\beta_i) - A_i^{(m)}),\\ \frac{\partial Q(\boldsymbol{\theta}_G, \boldsymbol{\theta}_G^{(m)})}{\partial b_j} &= -\sum_{i=1}^{I} K_i x_{ij}\left(\alpha_i - \frac{B_i^{(m)}}{\beta_i}\right),\end{aligned}$$

for $j = 0, 1, \dots, J$. The corresponding second-order derivatives are

$$\begin{aligned}\frac{\partial^2 Q(\boldsymbol{\theta}_G, \boldsymbol{\theta}_G^{(m)})}{\partial a_p\,\partial a_q} &= -\sum_{i=1}^{I} K_i x_{ip} x_{iq}\alpha_i(\alpha_i \Psi'(\alpha_i) + \Psi(\alpha_i) + \ln(\beta_i) - A_i^{(m)}),\\ \frac{\partial^2 Q(\boldsymbol{\theta}_G, \boldsymbol{\theta}_G^{(m)})}{\partial b_p\,\partial b_q} &= -\sum_{i=1}^{I} \frac{K_i x_{ip} x_{iq} B_i^{(m)}}{\beta_i},\\ \frac{\partial^2 Q(\boldsymbol{\theta}_G, \boldsymbol{\theta}_G^{(m)})}{\partial a_p\,\partial b_q} &= -\sum_{i=1}^{I} K_i x_{ip} x_{iq}\alpha_i,\end{aligned}$$

for $p, q = 0, 1, \dots, J$, where $\Psi(z) = \partial \ln \Gamma(z)/\partial z$ and $\Psi'(z)$ are the digamma and trigamma functions, respectively; see Abramowitz and Stegun (1972) for a discussion on the computation of these functions.

Here again, the conditional expectations required in the E-step, $A_i^{(m)}$ and $B_i^{(m)}$, only involve the failed and unfailed devices. For the failed devices, the proportion is n_i/K_i, and the lifetimes are right-truncated at the inspection time τ_i, while for the unfailed devices, the proportion is $1 - n_i/K_i$, and the lifetimes are left-truncated at the inspection time τ_i. Consequently, we obtain the conditional expectations, given the observed data and the current estimate $\boldsymbol{\theta}_G^{(m)}$ of $\boldsymbol{\theta}_G$, as follows:

$$
\begin{aligned}
A_i^{(m)} &= E\left[\ln\left(\frac{T_i}{\beta_i^{(m)}}\right)\middle|\mathbf{z}, \boldsymbol{\theta}_G^{(m)}\right] + \ln(\beta_i^{(m)}) \\
&= \ln(\beta_i^{(m)}) + \left(\frac{n_i}{K_i}\right)\frac{\int_0^{\tau_i} \ln\left(\frac{t}{\beta_i^{(m)}}\right) f_G(t; \boldsymbol{\theta}_G^{(m)})dt}{F_G(\tau_i; \boldsymbol{\theta}_G^{(m)})} \\
&\quad + \left(1 - \frac{n_i}{K_i}\right)\frac{\int_{\tau_i}^{\infty} \ln\left(\frac{t}{\beta_i^{(m)}}\right) f_G(t; \boldsymbol{\theta}_G^{(m)})dt}{1 - F_G(\tau_i; \boldsymbol{\theta}_G^{(m)})} \\
&= \ln(\beta_i^{(m)}) + \left(\frac{n_i}{K_i}\right)\frac{\int_0^{\tau_i/\beta_i^{(m)}} \ln(t)t^{\alpha_i^{(m)}-1}\exp(-t)dt}{\Gamma(\alpha_i^{(m)})F_G(\tau_i; \boldsymbol{\theta}_G^{(m)})} \\
&\quad + \left(1 - \frac{n_i}{K_i}\right)\frac{\int_{\tau_i/\beta_i^{(m)}}^{\infty} \ln(t)t^{\alpha_i^{(m)}-1}\exp(-t)dt}{\Gamma(\alpha_i^{(m)})(1 - F_G(\tau_i; \boldsymbol{\theta}_G^{(m)}))} \\
&= \ln(\beta_i^{(m)}) + H_1\left(\alpha_i^{(m)}, \frac{\tau_i}{\beta_i^{(m)}}\right)\left(\frac{\hat{F}(\tau_i)}{F_G(\tau_i; \boldsymbol{\theta}_G^{(m)})}\right) \\
&\quad + \left\{\Psi(\alpha_i^{(m)}) - H_1\left(\alpha_i^{(m)}, \frac{\tau_i}{\beta_i^{(m)}}\right)\right\}\left(\frac{1 - \hat{F}(\tau_i)}{1 - F_G(\tau_i; \boldsymbol{\theta}_G^{(m)})}\right)
\end{aligned}
$$

and

$$
\begin{aligned}
B_i^{(m)} &= E[T_i|\mathbf{z}, \boldsymbol{\theta}_G^{(m)}] \\
&= \left(\frac{n_i}{K_i}\right)\frac{\int_0^{\tau_i} t f_G(t; \alpha_i^{(m)}, \beta_i^{(m)})dt}{F_G(\tau_i; \boldsymbol{\theta}_G^{(m)})} + \left(1 - \frac{n_i}{K_i}\right)\frac{\int_{\tau_i}^{\infty} t f_G(t; \alpha_i^{(m)}, \beta_i^{(m)})dt}{1 - F_G(\tau_i; \boldsymbol{\theta}_G^{(m)})} \\
&= \alpha_i^{(m)}\beta_i^{(m)}\gamma\left(\alpha_i^{(m)} + 1, \frac{\tau_i}{\beta_i^{(m)}}\right)\left(\frac{\hat{F}(\tau_i)}{F_G(\tau_i; \boldsymbol{\theta}_G^{(m)})}\right) \\
&\quad + \alpha_i^{(m)}\beta_i^{(m)}\Gamma\left(\alpha_i^{(m)} + 1, \frac{\tau_i}{\beta_i^{(m)}}\right)\left(\frac{1 - \hat{F}(\tau_i)}{1 - F_G(\tau_i; \boldsymbol{\theta}_G^{(m)})}\right),
\end{aligned}
$$

where

$$H_1(a,b) = \frac{1}{\Gamma(a)}\int_0^b \ln(x)x^{a-1}\exp(-x)dx$$
$$= \ln(b)\gamma(a,b) - \frac{b^a{}_2F_2(a,a;a+1,a+1;-b)}{a^{2\Gamma(a)}},$$

and ${}_nF_m(a_1,a_2,\ldots,a_n;b_1,b_2,\ldots,b_m;z)$ is the generalized hypergeometric function; one may refer to Slater (2008) for properties and computational formulas for generalized hypergeometric functions. The generalized hypergeometric function is available in mathematical programs such as Matlab and Maple. Details on the derivation of the above expression of $H_1(a,b)$ are presented in Appendix A.

We then have the MLEs of mean lifetime and reliability at mission time t under normal operating conditions $\mathbf{x}_0 = (x_{01}, x_{02}, \ldots, x_{0J})$, using the MLEs of the model parameters $\hat{\boldsymbol{\theta}}_G = (\hat{a}_0, \hat{a}_1, \ldots, \hat{a}_J, \hat{b}_0, \hat{b}_1, \ldots, \hat{b}_J)$, to be

$$\hat{\mu}_G = \hat{\alpha}_0\hat{\beta}_0 = \exp\left(\sum_{j=0}^{J}(\hat{a}_j + \hat{b}_j)x_{0j}\right)$$

and

$$\hat{R}_G(t) = \Gamma\left(\hat{\alpha}_0, \frac{t}{\hat{\beta}_0}\right),$$

where, from (2.15),

$$\hat{\alpha}_0 = \exp\left(\sum_{j=0}^{J}\hat{a}_j x_{0j}\right) \quad \text{and} \quad \hat{\beta}_0 = \exp\left(\sum_{j=0}^{J}\hat{b}_j x_{0j}\right).$$

2.3.3 Weibull Distribution

Similar to gamma distribution, Weibull distribution can also model increasing, decreasing, and constant hazard rates, and it also includes exponential distribution as a special case; see, for example, Johnson et al. (1994) and Murthy et al. (2003) on various methods and applications of Weibull distribution. In this case, Balakrishnan and Ling (2013) studied one-shot device testing data for the Weibull lifetime distribution with both scale and shape parameters varying over multiple stress factors.

Let us use W to denote the Weibull distribution in what follows. Specifically, in the ith test group, we assume that the lifetimes of devices follow the Weibull distribution, with shape parameter $\eta_i > 0$ and scale parameter $\beta_i > 0$, with cdf and pdf as

$$F_W(t;\eta_i,\beta_i) = 1 - \exp\left(-\left(\frac{t}{\beta_i}\right)^{\eta_i}\right), \quad t > 0, \tag{2.17}$$

and

$$f_W(t; \eta_i, \beta_i) = \frac{\partial F_W(t; \eta_i, \beta_i)}{\partial t} = \frac{\eta_i}{\beta_i}\left(\frac{t}{\beta_i}\right)^{\eta_i - 1} \exp\left(-\left(\frac{t}{\beta_i}\right)^{\eta_i}\right), \quad t > 0, \quad (2.18)$$

respectively. The mean lifetime and reliability at mission time t under normal operating conditions $\mathbf{x}_0 = (x_{01}, x_{02}, \dots, x_{0J})$ are

$$\mu_W = \int_0^{\infty} t f_W(t; \eta_0, \beta_0) dt = \beta_0 \Gamma\left(1 + \frac{1}{\eta_0}\right) \quad (2.19)$$

and

$$R_W(t) = 1 - F_W(t; \eta_0, \beta_0) = \exp\left(-\left(\frac{t}{\beta_0}\right)^{\eta_0}\right), \quad (2.20)$$

respectively. For extrapolating the data observed from the elevated stress levels $\mathbf{x}_i = (x_{i1}, x_{i2}, \dots, x_{iJ})$ to the lifetime characteristics of devices under normal operating conditions $\mathbf{x}_0 = (x_{01}, x_{02}, \dots, x_{0J})$, in this case, both parameters η_i and β_i are related to the stress levels in log-linear link functions of the form, with $x_{i0} = 1$,

$$\ln(\eta_i) = \sum_{j=0}^{J} r_j x_{ij} \quad \text{and} \quad \ln(\beta_i) = \sum_{j=0}^{J} s_j x_{ij}. \quad (2.21)$$

Thus, the new model parameter of the considered model is

$$\theta_W = (r_0, r_1, \dots, r_J, s_0, s_1, \dots, s_J).$$

Instead of working with Weibull lifetimes for lifetime data, it is often more convenient to work with extreme value distribution for log-lifetimes. This is because the Weibull distribution belongs to the scale-shape family, while the extreme value distribution belongs to a location-scale family, which makes it convenient to work with; see Johnson et al. (1995). For this reason, we consider here the extreme value distribution with corresponding pdf and cdf as

$$f_V(v; \mu_i, \sigma_i) = \frac{\xi_i \exp(-\xi_i)}{\sigma_i}, -\infty < v < \infty, \quad (2.22)$$

and

$$F_V(v; \mu_i, \sigma_i) = 1 - \exp(-\xi_i), -\infty < v < \infty, \quad (2.23)$$

where $v = \ln(t)$, $\xi_i = \exp((v - \mu_i)/\sigma_i)$, with link functions $\mu_i = \ln(\beta_i) = \sum_{j=0}^{J} s_j x_{ij}$ and $\sigma_i = 1/\eta_i = \exp(-\sum_{j=0}^{J} r_j x_{ij})$.

We now define $v_{ik} = \ln(t_{ik})$ and $\xi_{ik} = \exp((v_{ik} - \mu_i)/\sigma_i)$. Then, under the EM framework, the log-likelihood function, based on the complete data, is obtained readily from (2.22) as

$$\ell_c = \sum_{i=1}^{I} \sum_{k=1}^{K_i} \ln(f_V(v_{ik}; \mu_i, \sigma_i))$$

$$= \sum_{i=1}^{I} \left(K_i \sum_{j=0}^{J} r_j x_{ij} + \sum_{k=1}^{K_i} \ln(\xi_{ik}) - \sum_{k=1}^{K_i} \xi_{ik} \right).$$

Let $A_i^{(m)} = E[\ln(\xi_i) | \mathbf{z}, \boldsymbol{\theta}_W^{(m)}]$ and $B_i^{(m)} = E[\xi_i | \mathbf{z}, \boldsymbol{\theta}_W^{(m)}]$. Then, the conditional expectation, from (2.3), is

$$Q(\boldsymbol{\theta}_W, \boldsymbol{\theta}_W^{(m)}) = \sum_{i=1}^{I} K_i \left(\sum_{j=0}^{J} r_j x_{ij} + A_i^{(m)} - B_i^{(m)} \right). \tag{2.24}$$

The first-order derivatives with respect to the model parameters, required for the maximization of the quantity $Q(\boldsymbol{\theta}_W, \boldsymbol{\theta}_W^{(m)})$ in (2.24), are given by

$$\frac{\partial Q(\boldsymbol{\theta}_W, \boldsymbol{\theta}_W^{(m)})}{\partial r_j} = \sum_{i=1}^{I} K_i x_{ij} (1 + A_i^{(m)} - C_i^{(m)}),$$

$$\frac{\partial Q(\boldsymbol{\theta}_W, \boldsymbol{\theta}_W^{(m)})}{\partial s_j} = -\sum_{i=1}^{I} \frac{K_i x_{ij} (1 - B_i^{(m)})}{\sigma_i},$$

for $j = 0, 1, \ldots, J$, where $C_i^{(m)} = E[\xi_i \ln(\xi_i) | \mathbf{z}, \boldsymbol{\theta}_W^{(m)}]$. The corresponding second-order derivatives are

$$\frac{\partial^2 Q(\boldsymbol{\theta}_W, \boldsymbol{\theta}_W^{(m)})}{\partial r_p \, \partial r_q} = \sum_{i=1}^{I} K_i x_{ip} x_{iq} (A_i^{(m)} - C_i^{(m)} - D_i^{(m)}),$$

$$\frac{\partial^2 Q(\boldsymbol{\theta}_W, \boldsymbol{\theta}_W^{(m)})}{\partial s_p \, \partial s_q} = -\sum_{i=1}^{I} \frac{K_i x_{ip} x_{iq} B_i^{(m)}}{\sigma_i^2},$$

$$\frac{\partial^2 Q(\boldsymbol{\theta}_W, \boldsymbol{\theta}_W^{(m)})}{\partial r_p \, \partial s_q} = -\sum_{i=1}^{I} \frac{K_i x_{ip} x_{iq} (1 - B_i^{(m)} - C_i^{(m)})}{\sigma_i},$$

for $p, q = 0, 1, \ldots, J$, where $D_i^{(m)} = E[\xi_i (\ln(\xi_i))^2 | \mathbf{z}, \boldsymbol{\theta}_W^{(m)}]$.

The required conditional expectations in the E-step, namely, $A_i^{(m)}, B_i^{(m)}, C_i^{(m)}$ and $D_i^{(m)}$, again involve only the failed and unfailed devices. For the failed devices, the proportion is n_i/K_i and the lifetimes are right-truncated at the inspection time τ_i, while for the unfailed devices, the proportion is $1 - n_i/K_i$ and the lifetimes are left-truncated at the inspection time τ_i. Upon defining $v_i^{(m)} = \exp((\ln(\tau_i) - \mu_i^{(m)})/\sigma_i^{(m)})$, we can derive the required conditional expectations to be as follows:

$$\begin{aligned} A_i^{(m)} &= E[\ln(\xi_i) | \mathbf{z}, \boldsymbol{\theta}_W^{(m)}] \\ &= \left(\frac{n_i}{K_i} \right) \frac{\int_0^{\ln(\tau_i)} \ln(\xi) f_V(v; \boldsymbol{\theta}_W^{(m)}) dv}{F_V(\ln(\tau_i); \boldsymbol{\theta}_W^{(m)})} + \left(1 - \frac{n_i}{K_i} \right) \frac{\int_{\ln(\tau_i)}^{\infty} \ln(\xi) f_V(v; \boldsymbol{\theta}_W^{(m)}) dv}{1 - F_V(\ln(\tau_i); \boldsymbol{\theta}_W^{(m)})} \\ &= \left(\frac{n_i}{K_i} \right) \frac{\int_0^{v_i^{(m)}} \ln(x) \exp(-x) dx}{F_W(\tau_i; \boldsymbol{\theta}_W^{(m)})} + \left(1 - \frac{n_i}{K_i} \right) \frac{\int_{v_i^{(m)}}^{\infty} \ln(x) \exp(-x) dx}{1 - F_W(\tau_i; \boldsymbol{\theta}_W^{(m)})} \end{aligned}$$

$$= -\left\{\gamma + \frac{\ln(v_i^{(m)})}{\exp(v_i^{(m)})} + \mathrm{EI}(v_i^{(m)})\right\}\left(\frac{\hat{F}(\tau_i)}{F_W(\tau_i;\boldsymbol{\theta}_W^{(m)})}\right)$$

$$+ \left\{\frac{\ln(v_i^{(m)})}{\exp(v_i^{(m)})} + \mathrm{EI}(v_i^{(m)})\right\}\left(\frac{1-\hat{F}(\tau_i)}{1-F_W(\tau_i;\boldsymbol{\theta}_W^{(m)})}\right),$$

$$B_i^{(m)} = E[\xi_i|\mathbf{z},\boldsymbol{\theta}_W^{(m)}]$$

$$= \left(\frac{n_i}{K_i}\right)\frac{\int_0^{\ln(\tau_i)} \xi f_V(v;\boldsymbol{\theta}_W^{(m)})dv}{F_V(\ln(\tau_i);\boldsymbol{\theta}_W^{(m)})} + \left(1-\frac{n_i}{K_i}\right)\frac{\int_{\ln(\tau_i)}^{\infty} \xi f_V(v;\boldsymbol{\theta}_W^{(m)})dv}{1-F_V(\ln(\tau_i);\boldsymbol{\theta}_W^{(m)})}$$

$$= \left(\frac{n_i}{K_i}\right)\frac{\int_0^{v_i^{(m)}} x\exp(-x)dx}{F_W(\tau_i;\boldsymbol{\theta}_W^{(m)})} + \left(1-\frac{n_i}{K_i}\right)\frac{\int_{v_i^{(m)}}^{\infty} x\exp(-x)dx}{1-F_W(\tau_i;\boldsymbol{\theta}_W^{(m)})}$$

$$= \left\{1-\frac{1+v_i^{(m)}}{\exp(v_i^{(m)})}\right\}\left(\frac{\hat{F}(\tau_i)}{F_W(\tau_i;\boldsymbol{\theta}_W^{(m)})}\right)$$

$$+ \left\{\frac{1+v_i^{(m)}}{\exp(v_i^{(m)})}\right\}\left(\frac{1-\hat{F}(\tau_i)}{1-F_W(\tau_i;\boldsymbol{\theta}_W^{(m)})}\right),$$

$$C_i^{(m)} = E[\xi_i\ln(\xi_i)|\mathbf{z},\boldsymbol{\theta}_W^{(m)}]$$

$$= \left(\frac{n_i}{K_i}\right)\frac{\int_0^{\ln(\tau_i)} \xi\ln(\xi) f_V(v;\boldsymbol{\theta}_W^{(m)})dv}{F_V(\ln(\tau_i);\boldsymbol{\theta}_W^{(m)})}$$

$$+ \left(1-\frac{n_i}{K_i}\right)\frac{\int_{\ln(\tau_i)}^{\infty} \xi\ln(\xi) f_V(v;\boldsymbol{\theta}_W^{(m)})dv}{1-F_V(\ln(\tau_i);\boldsymbol{\theta}_W^{(m)})}$$

$$= \left(\frac{n_i}{K_i}\right)\frac{\int_0^{v_i^{(m)}} x\ln(x)\exp(-x)dx}{F_W(\tau_i;\boldsymbol{\theta}_W^{(m)})} + \left(1-\frac{n_i}{K_i}\right)\frac{\int_{v_i^{(m)}}^{\infty} x\ln(x)\exp(-x)dx}{1-F_W(\tau_i;\boldsymbol{\theta}_W^{(m)})}$$

$$= \left\{1-\gamma-\mathrm{EI}(v_i^{(m)}) - \frac{1+\ln(v_i^{(m)})+v_i^{(m)}\ln(v_i^{(m)})}{\exp(v_i^{(m)})}\right\}\left(\frac{\hat{F}(\tau_i)}{F_W(\tau_i;\boldsymbol{\theta}_W^{(m)})}\right)$$

$$+ \left\{\mathrm{EI}(v_i^{(m)}) + \frac{1+\ln(v_i^{(m)})+v_i^{(m)}\ln(v_i^{(m)})}{\exp(v_i^{(m)})}\right\}\left(\frac{1-\hat{F}(\tau_i)}{1-F_W(\tau_i;\boldsymbol{\theta}_W^{(m)})}\right),$$

$$D_i^{(m)} = E[\xi_i(\ln(\xi_i))^2|\mathbf{z},\boldsymbol{\theta}_W^{(m)}]$$

$$= \left(\frac{n_i}{K_i}\right)\frac{\int_0^{\ln(\tau_i)} \xi(\ln(\xi))^2 f_V(v;\boldsymbol{\theta}_W^{(m)})dv}{F_V(\ln(\tau_i);\boldsymbol{\theta}_W^{(m)})}$$

$$
\begin{aligned}
&+\left(1-\frac{n_i}{K_i}\right)\frac{\int_{\ln(\tau_i)}^{\infty}\xi(\ln(\xi))^2 f_V(v;\boldsymbol{\theta}_W^{(m)})dv}{1-F_V(\ln(\tau_i);\boldsymbol{\theta}_W^{(m)})}\\
&=\left(\frac{n_i}{K_i}\right)\frac{\int_0^{v_i^{(m)}} x(\ln(x))^2\exp(-x)dx}{F_W(\tau_i;\boldsymbol{\theta}_W^{(m)})}\\
&+\left(1-\frac{n_i}{K_i}\right)\frac{\int_{v_i^{(m)}}^{\infty} x(\ln(x))^2\exp(-x)dx}{1-F_W(\tau_i;\boldsymbol{\theta}_W^{(m)})}\\
&=H_2(2,v_i^{(m)})\left(\frac{\hat{F}(\tau_i)}{F_W(\tau_i;\boldsymbol{\theta}_W^{(m)})}\right)\\
&+\left\{\gamma^2+\frac{\pi^2}{6}-2\gamma-H_2(2,v_i^{(m)})\right\}\left(\frac{1-\hat{F}(\tau_i)}{1-F_W(\tau_i;\boldsymbol{\theta}_W^{(m)})}\right),
\end{aligned}
$$

where

$$
\begin{aligned}
H_2(a,b)&=\frac{1}{\Gamma(a)}\int_0^b(\ln(x))^2x\exp(-x)dx\\
&=(\ln(b))^2\gamma(a,b)-\frac{2b^a\ln(b)_2F_2(a,a;a+1,a+1;-b)}{a^{2\Gamma(a)}}\\
&+\frac{2b^a{}_3F_3(a,a,a;a+1,a+1,a+1;-b)}{a^{3\Gamma(a)}},
\end{aligned}
$$

$\gamma \approx 0.577215665$ is the Euler's constant, $\mathrm{EI}(a)=\int_a^{\infty}x^{-1}\exp(-x)dx$ is the exponential integral, and ${}_nF_m(a_1,a_2,\ldots,a_n;b_1,b_2,\ldots,b_m;z)$ is the generalized hypergeometric function. The exponential integral is available in mathematical programs such as R, Matlab and Maple. Details on the derivation of the above expression of $H_2(a,b)$ are presented in Appendix A.

We then have the MLEs of mean lifetime and reliability at mission time t under normal operating conditions $\mathbf{x}_0=(x_{01},x_{02},\ldots,x_{0J})$, when the MLEs of the model parameters $\hat{\boldsymbol{\theta}}_W=(\hat{r}_0,\hat{r}_1,\ldots,\hat{r}_J,\hat{s}_0,\hat{s}_1,\ldots,\hat{s}_J)$ have been determined, from (2.19) and (2.20), to be

$$
\hat{\mu}_W=\hat{\beta}_0\Gamma\left(1+\frac{1}{\hat{\eta}_0}\right)
$$

and

$$
\hat{R}_W(t)=\exp\left[-\left(\frac{t}{\hat{\beta}_0}\right)^{\hat{\eta}_0}\right],
$$

where

$$
\hat{\eta}_0=\exp\left(\sum_{j=0}^{J}\hat{r}_jx_{0j}\right)\quad\text{and}\quad\hat{\beta}_0=\exp\left(\sum_{j=0}^{J}\hat{s}_jx_{0j}\right).
$$

2.4 Interval Estimation

2.4.1 Asymptotic Confidence Intervals

To construct confidence intervals for parameters of interest, asymptotic confidence intervals are commonly used, in conjunction with MLEs, when sample sizes are sufficiently large. This would require the asymptotic variance-covariance matrix of the MLEs of model parameters, which could be obtained as the inverse of the observed information matrix.

Under the EM framework, the missing information principle (Louis, 1982) is often employed for obtaining the observed information matrix of the MLEs of model parameters. In the case of one-shot device testing data, if $\hat{\boldsymbol{\theta}}$ is the maximum likelihood estimate of $\boldsymbol{\theta}$, the observed information matrix obtained from the missing information principle is equivalent to the observed Fisher information matrix computed directly as

$$I_{\text{obs}}(\boldsymbol{\theta}) = \sum_{i=1}^{I} K_i \left(\frac{1}{F(\tau_i;\boldsymbol{\theta})} + \frac{1}{1 - F(\tau_i;\boldsymbol{\theta})} \right) \left(\frac{\partial F(\tau_i;\boldsymbol{\theta})}{\partial \boldsymbol{\theta}} \right) \left(\frac{\partial F(\tau_i;\boldsymbol{\theta})}{\partial \boldsymbol{\theta}'} \right). \quad (2.25)$$

Details of derivation of the observed information matrix in (2.25) are presented in Appendix B.

The first-order derivatives of the cdf, $F(\tau_i;\boldsymbol{\theta})$, under the exponential, gamma, and Weibull lifetime distributions, with respect to the corresponding model parameters (exponential: $\boldsymbol{\theta}_E = (e_j)$, gamma: $\boldsymbol{\theta}_G = (a_j, b_j)$, Weibull: $\boldsymbol{\theta}_W = (r_j, s_j)$ for $j = 0, 1, \ldots, J$), can be found to be as follows:

$$\frac{\partial F_E(\tau_i;\boldsymbol{\theta}_E)}{\partial e_j} = -\tau_i \lambda_i x_{ij} \exp(-\tau_i \lambda_i), \quad (2.26)$$

$$\frac{\partial F_G(\tau_i;\boldsymbol{\theta}_G)}{\partial a_j} = -\alpha_i x_{ij} \left(\Psi(\alpha_i)\gamma\left(\alpha_i, \frac{\tau_i}{\beta_i}\right) - H_1\left(\alpha_i, \frac{\tau_i}{\beta_i}\right) \right), \quad (2.27)$$

$$\frac{\partial F_G(\tau_i;\boldsymbol{\theta}_G)}{\partial b_j} = -\frac{x_{ij}}{\Gamma(\alpha_i)} \left(\frac{\tau_i}{\beta_i} \right)^{\alpha_i} \exp\left(-\frac{\tau_i}{\beta_i} \right), \quad (2.28)$$

$$\frac{\partial F_W(\tau_i;\boldsymbol{\theta}_W)}{\partial r_j} = x_{ij} \ln(v_i) v_i \exp(-v_i), \quad (2.29)$$

$$\frac{\partial F_W(\tau_i;\boldsymbol{\theta}_W)}{\partial s_j} = -\frac{x_{ij}}{\sigma_i} v_i \exp(-v_i). \quad (2.30)$$

Then, the asymptotic variance-covariance matrix of the MLEs of model parameters is the inverse of the observed information matrix, i.e.

$$V_{\boldsymbol{\theta}} = I_{\text{obs}}^{-1}(\boldsymbol{\theta}).$$

Furthermore, as mentioned earlier in the beginning of Section 2.3, the asymptotic variance of the MLE of a function of model parameter $\boldsymbol{\theta}$, such as mean

Table 2.2 The first-order derivatives of mean lifetime and reliability function under normal operating conditions $\mathbf{x}_0$[a)] with respect to model parameters under the exponential, gamma and Weibull distributions.

	Mean lifetime μ	
Model	**Model parameter**	**The first-order derivative**
Exponential	e_j	$-\frac{x_{0j}}{\lambda_0}$
Gamma	a_j	$x_{0j}\alpha_0\beta_0$
	b_j	$x_{0j}\alpha_0\beta_0$
Weibull	r_j	$-x_{0j}\sigma_0 \exp(\mu_0)\Psi(1+\sigma_0)\Gamma(1+\sigma_0)$
	s_j	$x_{0j} \exp(\mu_0)\Gamma(1+\sigma_0)$
	Reliability $R(t)$	
Model	**Model parameter**	**The first-order derivative**
Exponential	e_j	$t\lambda_0 x_{0j} \exp(-t\lambda_0)$
Gamma	a_j	$\alpha_0 x_{0j}\left(\Psi(\alpha_0)\gamma\left(\alpha_0, \frac{t}{\beta_0}\right) - H_1\left(\alpha_0, \frac{t}{\beta_0}\right)\right)$
	b_j	$\frac{x_{0j}}{\Gamma(\alpha_0)}\left(\frac{t}{\beta_0}\right)^{\alpha_0} \exp\left(-\frac{t}{\beta_0}\right)$
Weibull[b)]	r_j	$-x_{0j} \ln(\xi_0)\xi_0 \exp(-\xi_0)$
	s_j	$\frac{x_{0j}}{\sigma_0}\xi_0 \exp(-\xi_0)$

a) $\mathrm{x}_0 = (x_{01}, x_{02}, \dots, x_{0J},)$
b) $\xi_0 = \exp((\ln(t) - \mu_0)/\sigma_0)$.

lifetime and reliability at mission time under normal operating conditions, can be readily obtained by delta method; this would require the first-order derivatives of the corresponding function with respect to the model parameters. The first-order derivatives of mean lifetime and reliability function for the exponential, gamma, and Weibull lifetime distributions, with respect to the corresponding model parameters, are all presented in Table 2.2. Specifically, the asymptotic variance of the MLE of a parameter of interest, ϕ, is then

$$\hat{V}_\phi = P'\hat{V}_\theta P,$$

where $P = \partial\phi/\partial\boldsymbol{\theta}$ is a column vector of the first-order derivatives of ϕ with respect to the components of $\boldsymbol{\theta}$, and $\hat{V}_\theta = I_{\text{obs}}^{-1}(\hat{\boldsymbol{\theta}})$ is the inverse of the observed Fisher information matrix evaluated at the MLE of model parameter, $\hat{\boldsymbol{\theta}}$.

Subsequently, the standard error $\text{se}(\boldsymbol{\phi}) = \sqrt{\hat{V}_{\phi}}$ can be readily obtained, with which a $100(1-\delta)\%$ asymptotic confidence interval for the parameter ϕ can be obtained as

$$\left(\hat{\phi} - z_{\frac{\delta}{2}}\,\text{se}(\boldsymbol{\phi}),\quad \hat{\phi} + z_{\frac{\delta}{2}}\,\text{se}(\boldsymbol{\phi})\right),$$

where $z_{\delta/2}$ is the upper $\delta/2$ percentage point of the standard normal distribution. Truncation on the bounds of confidence intervals for reliability and mean lifetime of devices may be needed, as the reliability has to be between 0 and 1, while the mean lifetime has to be necessarily positive. But, such a truncation of confidence intervals may not be accurate.

2.4.2 Approximate Confidence Intervals

Hoyle (1973) discussed various transformations for developing suitable confidence intervals in case of estimators with skewed distributions and/or when parameters have bounded ranges. When the sample size is small, the MLEs of mean lifetime and reliability at mission time may not possess normal distributions, and consequently, the confidence intervals constructed directly by the asymptotic method may not maintain the nominal level of confidence. For a detailed and comprehensive review of all such transformations and their use in developing statistical intervals, we refer the readers to Meeker et al. (2017).

For variance stabilization, when the asymptotic distribution is over-dispersed, the hyperbolic transformation (Anscombe, 1948; Kim and Taylor, 1994) could be used effectively. As the reliability lies between 0 and 1, a hyperbolic arcsecant (arsech) transformation could be suitable for developing confidence intervals for reliability. In this arsech-approach, we assume that

$$\hat{f} = \ln\left(\frac{1 + \sqrt{1 - (\hat{R}(t))^2}}{\hat{R}(t)}\right) \tag{2.31}$$

is normally distributed. Then, by delta method, its standard error can be readily found to be

$$\text{se}(\hat{f}) = \frac{\text{se}(\hat{R}(t))}{\hat{R}(t)\sqrt{1 - (\hat{R}(t))^2}},$$

where $\text{se}(\hat{R}(t))$ is the standard error of $\hat{R}(t)$. Thence, an approximate $100(1-\delta)\%$ confidence interval for f is given by

$$\left(\hat{f} - z_{\frac{\delta}{2}}\,\text{se}(\hat{f}),\quad \hat{f} + z_{\frac{\delta}{2}}\,\text{se}(\hat{f})\right).$$

As $\hat{f} \geq 0$, the lower bound may have to be truncated at 0. Since $\hat{f}$ in (2.31) is a monotone decreasing function of $\hat{R}(t)$, we can immediately obtain an approximate $100(1-\delta)\%$ confidence interval for the reliability $R(t)$ to be

$$\left(\frac{2}{\exp(U_f)+\exp(-U_f)}, \quad \frac{2}{\exp(L_f)+\exp(-L_f)}\right),$$

where $U_f = \hat{f} + z_{\frac{\delta}{2}}\,\text{se}(\hat{f})$ and $L_f = \max\left(0, \hat{f} - z_{\frac{\delta}{2}}\,\text{se}(\hat{f})\right)$.

The logit transformation (Viveros and Balakrishnan, 1993) is another useful transformation for this purpose. In this case, it is assumed that

$$\hat{f} = \ln\left(\frac{\hat{R}(t)}{1-\hat{R}(t)}\right) \tag{2.32}$$

is normally distributed with the corresponding standard error, determined by delta method, as

$$\text{se}(\hat{f}) = \frac{\text{se}(\hat{R}(t))}{\hat{R}(t)(1-\hat{R}(t))}.$$

Then, an approximate $100(1-\delta)\%$ confidence interval for f is

$$\left(\hat{f} - z_{\frac{\delta}{2}}\,\text{se}(\hat{f}), \quad \hat{f} + z_{\frac{\delta}{2}}\,\text{se}(\hat{f})\right),$$

using which we can obtain an approximate $100(1-\delta)\%$ confidence interval for the reliability $R(t)$, based on the inverse of transformation in (2.32), to be

$$\left(\frac{\hat{R}(t)}{\hat{R}(t)+(1-\hat{R}(t))S}, \quad \frac{\hat{R}(t)}{\hat{R}(t)+(1-\hat{R}(t))S^{-1}}\right),$$

where $S = \exp(z_{\frac{\delta}{2}}\,\text{se}(\hat{f}))$.

In a similar manner, the log-approach can be used to construct confidence intervals for mean lifetime, which would avoid having a negative lower bound for the mean lifetime. In this approach, we will assume that $\ln(\hat{\mu})$ is asymptotically normally distributed with the corresponding standard error as

$$\text{se}(\ln(\hat{\mu})) = \frac{\text{se}(\hat{\mu})}{\hat{\mu}},$$

where $\text{se}(\hat{\mu})$ is the standard error of $\hat{\mu}$. This approach yields readily an approximate $100(1-\delta)\%$ confidence interval for mean lifetime μ as

$$\left(\hat{\mu}\exp\left\{-\frac{z_{\frac{\delta}{2}}\,\text{se}(\hat{\mu})}{\hat{\mu}}\right\}, \quad \hat{\mu}\exp\left\{\frac{z_{\frac{\delta}{2}}\,\text{se}(\hat{\mu})}{\hat{\mu}}\right\}\right).$$

The computation of the required standard errors of the MLEs of mean lifetime and reliability at mission time have been discussed in the preceding subsection.

2.5 Simulation Studies

In this section, extensive simulation studies are carried out to evaluate the performance of the described inferential methods for various sample sizes and different lifetime distributions, in terms of bias, root mean square errors (RMSEs), coverage probabilities, and average widths of 95% confidence intervals, based on 1000 Monte Carlo simulations. Table 2.3 presents the CSALTs, with $I = 12$ test groups and $K_i = K$, for all i, used in these simulation studies.

The EM algorithm was terminated when the convergence was achieved by the criterion $||\theta^{(m)} - \theta^{(m+1)}|| \leq 1 \times 10^{-5}$. It was observed that the number of iterations, m, was at most 4015. The average numbers of iterations for different sample sizes and for different distributions were found, and these are presented in Table 2.4. It was also observed that the MLEs were obtained with less number of iterations through the EM algorithm when sample size increased. Finally, we should mention that the computational efforts for estimating the model parameters of Weibull distribution was more than those for exponential and gamma distributions.

Tables 2.5 and 2.6 present the bias and RMSEs of the MLEs of the model parameters, mean lifetime, and reliability under normal operating conditions for the exponential, gamma, and Weibull lifetime distributions. It can be observed that the

Table 2.3 CSALTs with two accelerating factors and $I = 12$ test groups.

Test group (i)	Stress levels (x_{i1}, x_{i2})	Inspection time (τ_i)
1	(30, 40)	10
2	(40, 40)	10
3	(30, 50)	10
4	(40, 50)	10
5	(30, 40)	20
6	(40, 40)	20
7	(30, 50)	20
8	(40, 50)	20
9	(30, 40)	30
10	(40, 40)	30
11	(30, 50)	30
12	(40, 50)	30

Table 2.4 Average numbers of iterations for different sample sizes under the exponential, gamma, and Weibull distributions.

Model	$K = 10$	$K = 30$	$K = 50$	$K = 100$	$K = 200$
Exponential	27.827	2325.08	2324.216	23.347	—
Gamma	—	159.871	142.493	129.187	120.696
Weibull	—	409.398	360.299	305.062	278.402

Table 2.5 Bias and RMSEs of MLEs of the model parameter, θ_E, mean lifetime, μ_E, and reliability, $R_E(t)$, under normal operating conditions $\mathbf{x}_0 = (25, 35)$, for various sample sizes, K, under exponential lifetime distribution.

		e_0	e_1	e_2	μ_E	$R_E(10)$	$R_E(20)$	$R_E(30)$
K		**−6.000**	**0.030**	**0.030**	**66.686**	**0.799**	**0.638**	**0.509**
10	Bias	0.003	0.000	−0.001	8.800	−0.013	−0.013	−0.007
	RMSE	1.760	0.031	0.030	38.985	0.090	0.134	0.154
30	Bias	0.000	0.000	0.000	2.512	−0.005	−0.006	−0.004
	RMSE	0.998	0.016	0.017	19.444	0.050	0.078	0.091
50	Bias	−0.012	0.001	0.000	1.806	−0.002	−0.002	−0.001
	RMSE	0.769	0.013	0.013	14.503	0.038	0.059	0.070
100	Bias	0.002	0.000	0.000	0.828	−0.001	−0.001	−0.001
	RMSE	0.537	0.009	0.009	10.043	0.026	0.042	0.050

bias of the parameters of interest are all small, except for the mean lifetime under the Weibull distribution in case of small sample sizes. In addition, the RMSEs are seen to decrease with increasing sample size. Further, the results show that the estimation of mean lifetime with sample size K less than 100 is not satisfactory for the Weibull distribution.

Tables 2.7–2.10 present the coverage probabilities and average widths of 95% confidence intervals for the model parameters, mean lifetime, and reliability under normal operating conditions for the exponential, gamma, and Weibull lifetime distributions.

From these tables, we observe that the asymptotic confidence intervals for the model parameters maintain the nominal level of confidence in all cases, but the coverage probabilities for mean lifetime and reliability are deflated (below the nominal level of confidence) for small sample sizes. This is so because the asymptotic distributions of the MLEs of mean lifetime and reliability become quite skewed, especially when the sample sizes are small. Figures 2.1 and 2.2 show that the normality assumption for mean lifetime and reliability is violated

Table 2.6 Bias and RMSEs of MLEs of the model parameters, $\{\theta_G, \theta_W\}$, mean lifetime, $\{\mu_G, \mu_W\}$, and reliability, $\{R_G(t), R_W(t)\}$, under normal operating conditions $\mathbf{x}_0 = (25, 35)$, for various sample sizes, K, under gamma, and Weibull lifetime distributions.

Gamma distribution

K		a_0 **−0.500**	a_1 **0.020**	a_2 **0.020**	b_0 **7.000**	b_1 **−0.060**	b_2 **−0.060**	μ_G **60.340**	$R_G(10)$ **0.912**	$R_G(20)$ **0.739**	$R_G(30)$ **0.562**
30	Bias	−0.101	0.002	0.001	0.187	−0.003	−0.002	4.049	−0.014	−0.010	−0.015
	RMSE	2.004	0.040	0.039	2.598	0.050	0.049	18.296	0.064	0.075	0.093
50	Bias	0.033	0.000	0.000	−0.002	0.000	0.000	1.875	−0.006	−0.003	−0.009
	RMSE	1.660	0.031	0.030	2.104	0.038	0.037	12.634	0.052	0.061	0.071
100	Bias	−0.021	0.000	0.000	0.061	0.000	−0.001	1.301	−0.003	−0.001	−0.002
	RMSE	1.141	0.021	0.022	1.446	0.026	0.026	8.563	0.036	0.043	0.048
200	Bias	0.000	0.000	0.001	0.010	0.001	−0.001	0.392	−0.001	−0.001	−0.002
	RMSE	0.851	0.015	0.016	1.068	0.018	0.019	5.757	0.027	0.031	0.034

Weibull distribution

K		r_0 **−1.000**	r_1 **0.020**	r_2 **0.020**	s_0 **6.000**	s_1 **−0.030**	s_2 **−0.030**	μ_W **62.451**	$R_W(10)$ **0.851**	$R_W(20)$ **0.686**	$R_W(30)$ **0.539**
30	Bias	−0.094	0.001	0.001	0.196	−0.002	−0.002	179.699	−0.009	−0.009	−0.022
	RMSE	1.746	0.031	0.032	1.386	0.020	0.024	2174.127	0.075	0.078	0.114
50	Bias	−0.016	0.001	0.000	0.125	−0.002	−0.001	18.066	−0.003	0.000	−0.009
	RMSE	1.323	0.024	0.024	0.947	0.016	0.016	84.059	0.061	0.058	0.086
100	Bias	−0.038	0.001	0.000	0.052	−0.001	0.000	5.611	−0.003	−0.002	−0.005
	RMSE	0.883	0.016	0.016	0.615	0.010	0.010	21.300	0.042	0.039	0.052
200	Bias	0.003	0.000	0.000	0.014	0.000	0.000	2.172	−0.001	−0.001	−0.003
	RMSE	0.608	0.012	0.011	0.428	0.007	0.007	12.136	0.030	0.028	0.036

Table 2.7 Coverage probabilities and average widths of 95% confidence intervals for the model parameter, θ_E, mean lifetime, μ_E, and reliability, $R_E(t)$, under normal operating conditions $\mathbf{x}_0 = (25, 35)$, for various sample sizes, K, under exponential lifetime distribution.

		e_0	e_1	e_2	μ_E	$R_E(10)$	$R_E(20)$	$R_E(30)$
K		**−6.000**	**0.030**	**0.030**	**66.686**	**0.799**	**0.638**	**0.509**
				Coverage probability				
10	asy[a)]	0.951	0.947	0.951	0.899	0.917	0.905	0.890
	arsech[b)]	—	—	—	—	0.945	0.941	0.937
	log/logit[c)]	—	—	—	0.943	0.958	0.966	0.969
30	asy[a)]	0.943	0.955	0.941	0.942	0.937	0.931	0.921
	arsech[b)]	—	—	—	—	0.941	0.946	0.951
	log/logit[c)]	—	—	—	0.936	0.939	0.940	0.945
50	asy[a)]	0.944	0.964	0.943	0.945	0.934	0.933	0.937
	arsech[b)]	—	—	—	—	0.939	0.938	0.937
	log/logit[c)]	—	—	—	0.951	0.953	0.951	0.948
100	asy[a)]	0.945	0.947	0.953	0.947	0.937	0.935	0.935
	arsech[b)]	—	—	—	—	0.943	0.941	0.943
	log/logit[c)]	—	—	—	0.940	0.939	0.938	0.940
				Average width				
10	asy[a)]	6.743	0.116	0.116	140.976	0.330	0.501	0.574
	arsech[b)]	—	—	—	—	0.312	0.474	0.563
	log/logit[c)]	—	—	—	163.664	0.326	0.466	0.527
30	asy[a)]	3.858	0.066	0.066	72.804	0.188	0.296	0.350
	arsech[b)]	—	—	—	—	0.185	0.289	0.345
	log/logit[c)]	—	—	—	76.227	0.188	0.288	0.336
50	asy[a)]	2.988	0.051	0.051	55.579	0.145	0.229	0.273
	arsech[b)]	—	—	—	—	0.143	0.226	0.271
	log/logit[c)]	—	—	—	57.121	0.145	0.226	0.267
100	asy[a)]	2.112	0.036	0.036	38.605	0.102	0.163	0.194
	arsech[b)]	—	—	—	—	0.102	0.162	0.194
	log/logit[c)]	—	—	—	39.134	0.103	0.162	0.192

a) asy: asymptotic confidence interval.
b) arsech: approximate confidence interval based on arsech transformation.
c) log/logit: approximate confidence interval based on log transformation for mean lifetime and logit transformation for reliability.

Table 2.8 Coverage probabilities and average widths of 95% asymptotic confidence intervals for the model parameters, $\{\theta_G, \theta_W\}$, for various sample sizes, K, under gamma and Weibull lifetime distributions.

Gamma distribution						
	a_0	a_1	a_2	b_0	b_1	b_2
	−0.500	0.020	0.020	7.000	−0.060	−0.060
Coverage probability						
$K = 30$	0.966	0.953	0.958	0.969	0.956	0.961
$K = 50$	0.945	0.954	0.958	0.946	0.957	0.956
$K = 100$	0.958	0.956	0.941	0.959	0.957	0.941
$K = 200$	0.940	0.944	0.946	0.938	0.947	0.950
Average width						
$K = 30$	8.483	0.155	0.154	10.827	0.190	0.190
$K = 50$	6.534	0.119	0.119	8.254	0.146	0.146
$K = 100$	4.585	0.084	0.084	5.783	0.102	0.103
$K = 200$	3.235	0.059	0.059	4.068	0.072	0.072
Weibull distribution						
	r_0	r_1	r_2	s_0	s_1	s_2
	−1.000	0.020	0.020	6.000	−0.030	−0.030
Coverage probability						
$K = 30$	0.947	0.954	0.958	0.946	0.954	0.938
$K = 50$	0.938	0.947	0.938	0.944	0.952	0.942
$K = 100$	0.952	0.949	0.951	0.948	0.956	0.947
$K = 200$	0.949	0.948	0.954	0.945	0.957	0.955
Average width						
$K = 30$	6.414	0.116	0.116	4.884	0.076	0.078
$K = 50$	4.878	0.089	0.089	3.522	0.057	0.056
$K = 100$	3.390	0.063	0.063	2.358	0.039	0.039
$K = 200$	2.383	0.044	0.044	1.632	0.027	0.027

Table 2.9 Coverage probabilities and average widths of 95% confidence intervals for mean lifetime, μ_G, and reliability, $R_G(t)$, under normal operating conditions $\mathbf{x}_0 = (25, 35)$, for various sample sizes, K, under gamma lifetime distribution.

		μ_G	$R_G(10)$	$R_G(20)$	$R_G(30)$
		60.340	**0.912**	**0.739**	**0.562**
		Coverage probability			
$K = 30$	asy[a]	0.935	0.923	0.947	0.958
	arsech[b]	—	0.959	0.960	0.974
	log/logit[c]	0.947	0.960	0.959	0.975
$K = 50$	asy[a]	0.942	0.906	0.943	0.950
	arsech[b]	—	0.933	0.954	0.958
	log/logit[c]	0.946	0.958	0.951	0.964
$K = 100$	asy[a]	0.946	0.935	0.953	0.955
	arsech[b]	—	0.955	0.956	0.956
	log/logit[c]	0.948	0.962	0.954	0.964
$K = 200$	asy[a]	0.945	0.922	0.942	0.953
	arsech[b]	—	0.938	0.946	0.954
	log/logit[c]	0.946	0.947	0.948	0.952
		Average width			
$K = 30$	asy[a]	68.361	0.232	0.313	0.363
	arsech[b]	—	0.245	0.302	0.357
	log/logit[c]	73.036	0.324	0.310	0.347
$K = 50$	asy[a]	47.479	0.187	0.241	0.279
	arsech[b]	—	0.191	0.236	0.276
	log/logit[c]	48.775	0.230	0.239	0.272
$K = 100$	asy[a]	32.311	0.143	0.170	0.190
	arsech[b]	—	0.141	0.168	0.189
	log/logit[c]	32.700	0.154	0.169	0.188
$K = 200$	asy[a]	22.015	0.102	0.120	0.133
	arsech[b]	—	0.101	0.120	0.133
	log/logit[c]	22.138	0.106	0.120	0.132

a) asy: asymptotic confidence interval.
b) arsech: approximate confidence interval based on arsech transformation.
c) log/logit: approximate confidence interval based on log transformation for mean lifetime and logit transformation for reliability.

Table 2.10 Coverage probabilities and average widths of 95% confidence intervals for mean lifetime, μ_W, and reliability, $R_W(t)$, under normal operating conditions $\mathbf{x}_0 = (25, 35)$, for various sample sizes, K, under Weibull lifetime distribution.

		μ_W 60.340	$R_W(10)$ 0.912	$R_W(20)$ 0.739	$R_W(30)$ 0.562
		Coverage probability			
$K = 30$	asy[a]	0.892	0.891	0.943	0.955
	arsech[b]	—	0.919	0.956	0.967
	log/logit[c]	0.927	0.941	0.953	0.981
$K = 50$	asy[a]	0.912	0.902	0.934	0.960
	arsech[b]	—	0.911	0.939	0.967
	log/logit[c]	0.930	0.932	0.941	0.979
$K = 100$	asy[a]	0.923	0.927	0.945	0.963
	arsech[b]	—	0.941	0.948	0.967
	log/logit[c]	0.938	0.941	0.949	0.971
$K = 200$	asy[a]	0.937	0.939	0.943	0.941
	arsech[b]	—	0.942	0.947	0.942
	log/logit[c]	0.949	0.942	0.947	0.944
		Average width			
$K = 30$	asy[a]	2.683e3	0.271	0.290	0.393
	arsech[b]	—	0.268	0.282	0.399
	log/logit[c]	1.843e11	0.304	0.283	0.380
$K = 50$	asy[a]	157.986	0.221	0.220	0.298
	arsech[b]	—	0.215	0.216	0.299
	log/logit[c]	512.418	0.232	0.217	0.290
$K = 100$	asy[a]	73.205	0.162	0.153	0.200
	arsech[b]	—	0.159	0.151	0.199
	log/logit[c]	78.626	0.165	0.152	0.197
$K = 200$	asy[a]	44.178	0.115	0.108	0.138
	arsech[b]	—	0.114	0.107	0.138
	log/logit[c]	45.175	0.116	0.107	0.137

a) asy: asymptotic confidence interval.
b) arsech: approximate confidence interval based on arsech transformation.
c) log/logit: approximate confidence interval based on log transformation for mean lifetime and logit transformation for reliability.

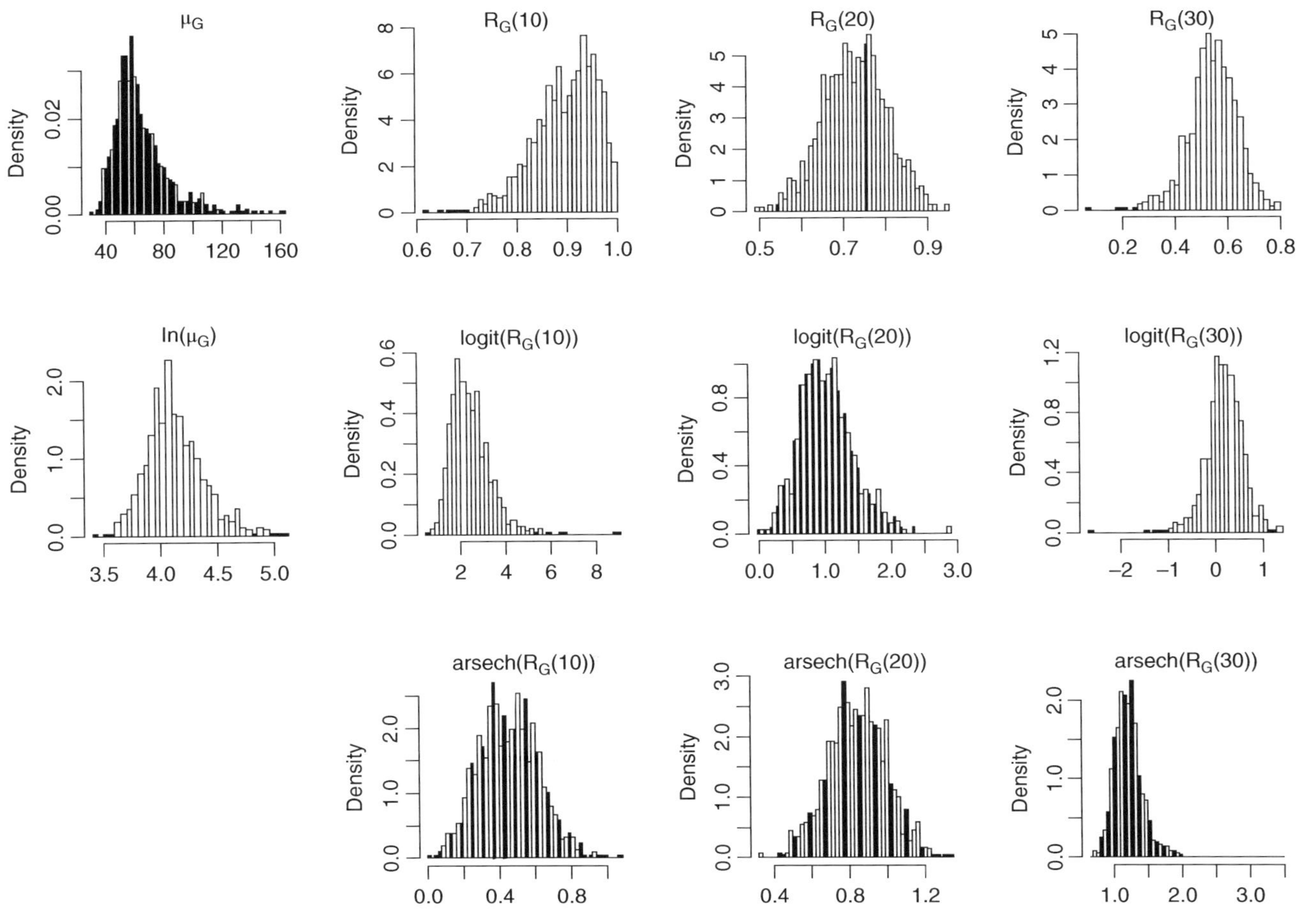

Figure 2.1 Histograms of the MLEs (first row), along with log/logit (second row) and arsech (third row) transformations, of mean lifetime and reliability, under gamma lifetime distribution for sample size $K = 30$.

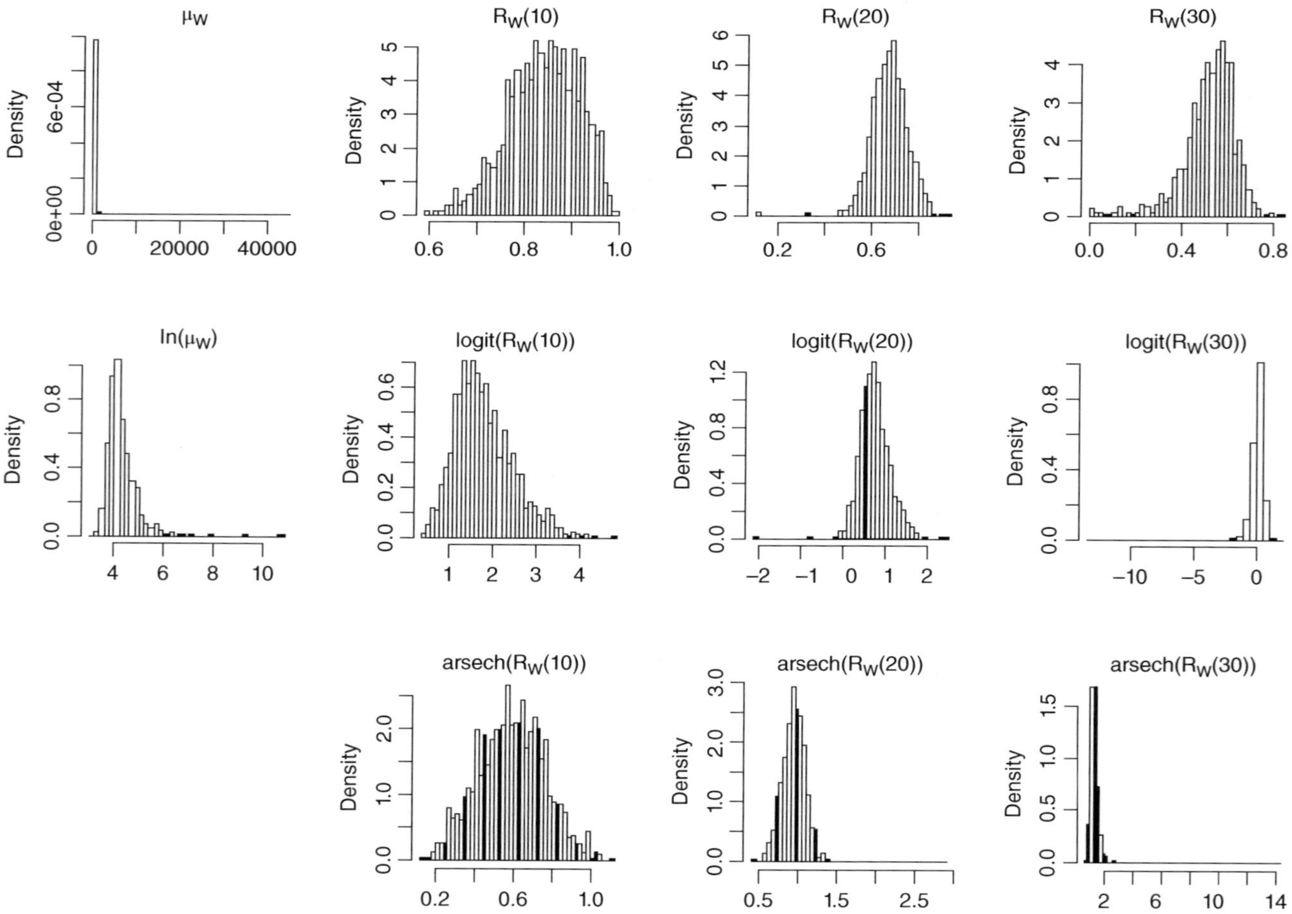

Figure 2.2 Histograms of the MLEs (first row), along with log/logit (second row) and arsech (third row) transformations, of mean lifetime and reliability under Weibull lifetime distribution for sample size $K = 30$.

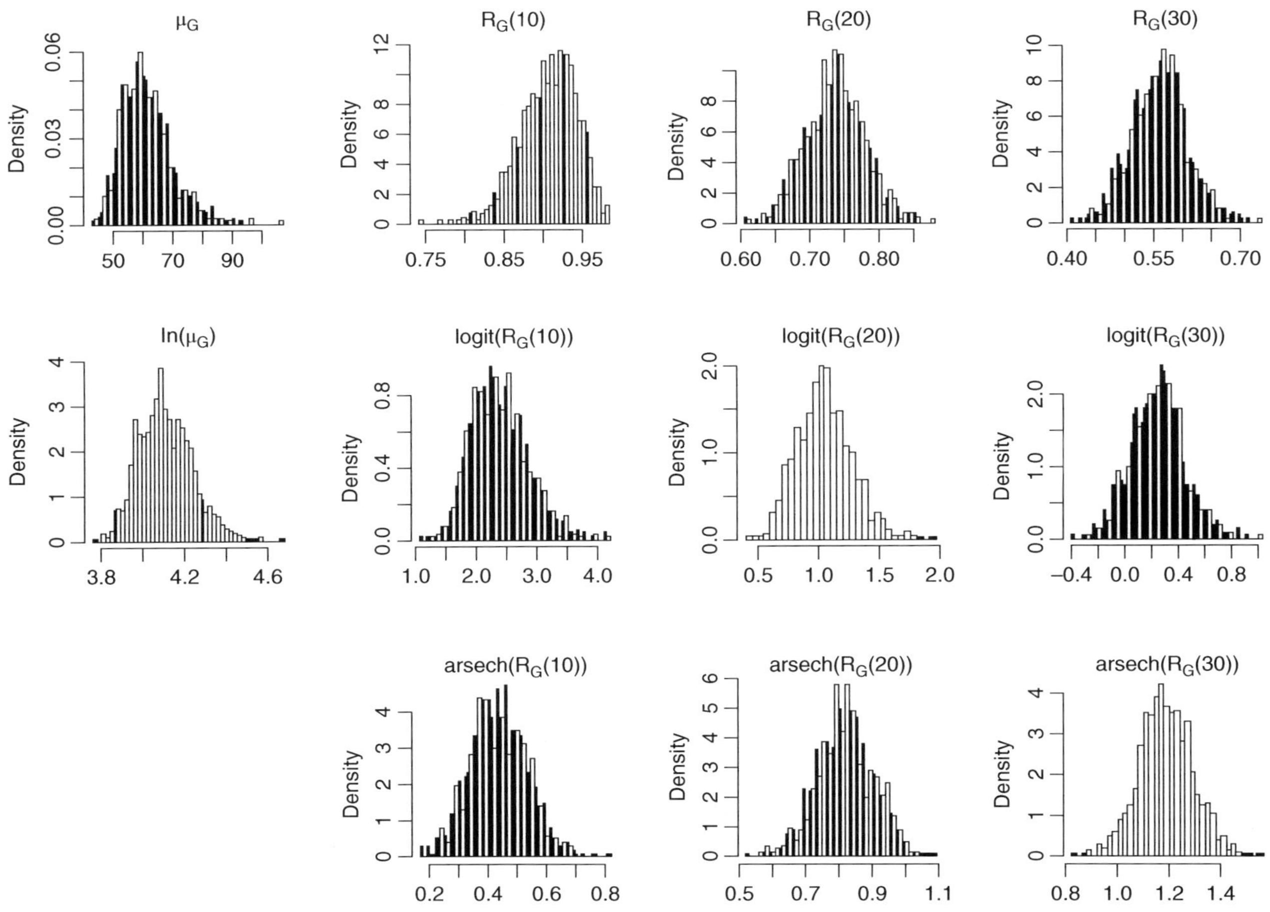

Figure 2.3 Histograms of the MLEs (first row), along with log/logit (second row) and arsech (third row) transformations, of mean lifetime and reliability, under gamma lifetime distribution for sample size $K = 100$.

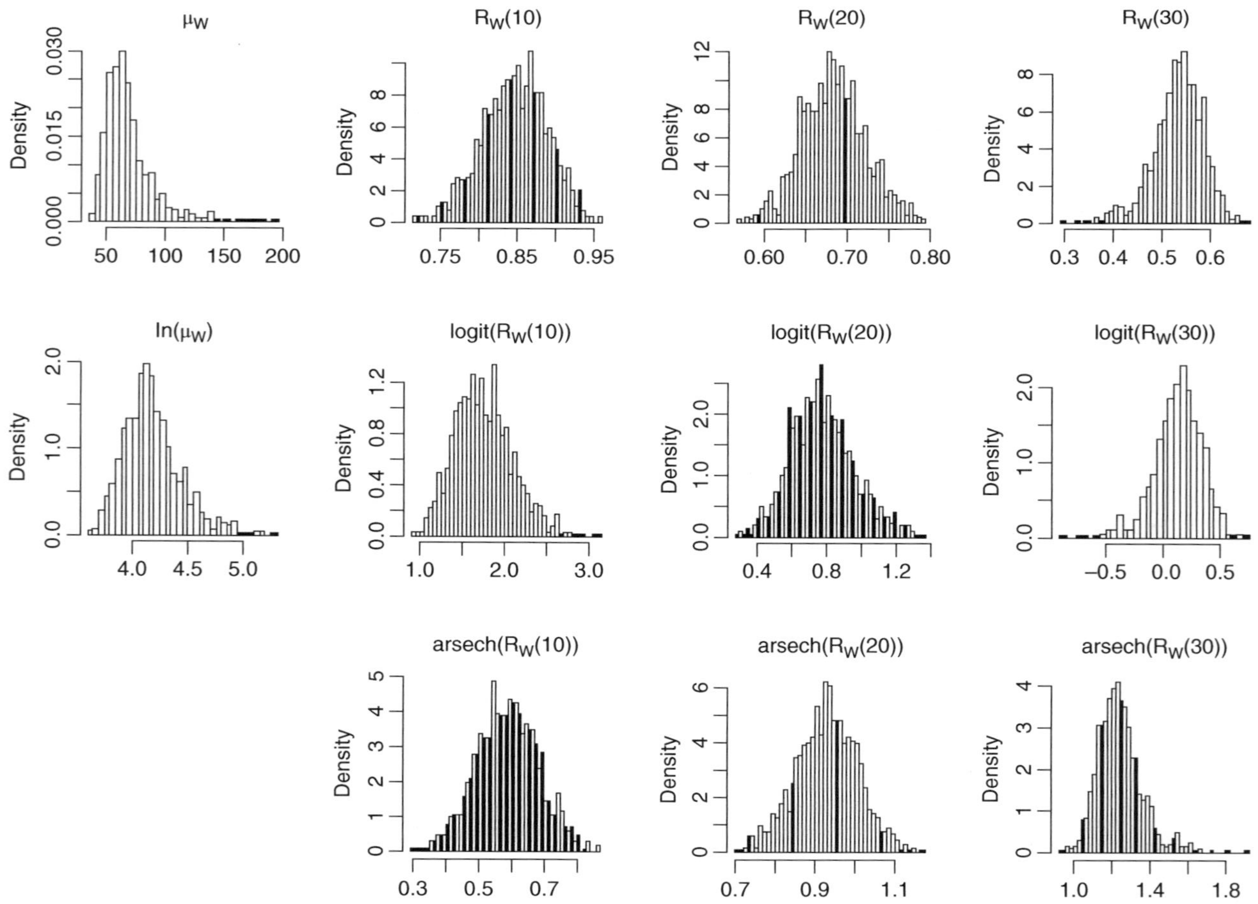

Figure 2.4 Histograms of the MLEs (first row), along with log/logit (second row) and arsech (third row) transformations, of mean lifetime and reliability, under Weibull lifetime distribution for sample size $K = 100$.

in the case of small sample sizes. On the other hand, the transformed MLEs possess approximate normal distributions, thus yielding satisfactory confidence intervals. We, therefore, observe that the coverage probabilities for mean lifetime and reliability can be improved significantly by using the approximate confidence intervals based on log- or arsech- or logit-transformation. From Figures 2.3 and 2.4, we also observe that all the distributions of the MLEs possess approximate normality in case of large sample sizes, and so all the confidence intervals have satisfactory coverage probabilities when the sample sizes are large. Moreover, these approximate confidence intervals also have average widths that are comparable to the asymptotic confidence intervals in all cases, except in the Weibull case for small sample sizes.

2.6 Case Studies with R Codes

In this section, we consider the one-shot device testing data under CSALTs presented in Tables 1.1 and 1.2. The R codes for defining the data, mission times, and normal operating conditions are displayed in Table 2.11. For the second data presented in Table 1.2, it is worth noting that the predictors, $x_1 = 1/\text{Temp}$ and

Table 2.11 R codes for defining the data in Tables 1.1 and 1.2.

```
# Electro-Explosive Devices Data
Temp<-rep(c(308,318,328),each=3) # temperatures (K)
X<-1/Temp
Tau<-rep(c(10,20,30),times=3) # inspection times
K<-rep(10,times=9) # sample sizes
n<-c(3,3,7,1,5,7,6,7,9) # number of failures
x0<-1/298 # normal operating conditions
t0<-c(10,20,30) # mission times
```

```
# Glass Capacitors Data
Temp<-rep(c(443,453),times=4) # temperatures (K)
Volt<-rep(seq(200,350,50),each=2) # voltages (volt)
X<-cbind(1/Temp,log(Volt)) #stress levels
Tau<-rep(c(450,400,350,300),times=2) # inspection times
K<-rep(8,times=8) # sample sizes
n<-c(1,0,0,1,3,4,3,2) # number of failures
x0<-c(1/443,log(250)) # normal operating conditions
t0<-c(500,800,1000) # mission times
```

Table 2.12 R codes for point estimation and 95% confidence intervals for the model parameters, mean lifetime, and reliability at mission times under normal operating conditions, for exponential, gamma, and Weibull lifetime distributions.

```
source("EM.R")
#exponential lifetime distribution
p<-n/K
p[p==0]<-0.01
p[p==1]<-0.99
y<-log(-log(1-p))-log(Tau)
A<-matrix(c(rep(1,each=length(Tau)),X),nrow=length(Tau))
thetaE0<-solve(t(A)%*%A,t(A)%*%y)
M<-EM(K,n,X,Tau,thetaE0,x0,t0,"exp")
AsyCI<-Asy(K,n,X,Tau,M$theta.hat,0.95,"exp")
AppCI<-Asy.phi(x0,t0,M$theta.hat,AsyCI$V,0.95,"exp")
#gamma lifetime distribution
theta0<-c(rep(0,length(thetaE0)),-thetaE0)
M<-EM(K,n,X,Tau,theta0,x0,t0,"gamma")
AsyCI<-Asy(K,n,X,Tau,M$theta.hat,0.95,"gamma")
AppCI<-Asy.phi(x0,t0,M$theta.hat,AsyCI$V,0.95,"gamma")
#Weibull lifetime distribution
M<-EM(K,n,X,Tau,theta0,x0,t0,"Weibull")
AsyCI<-Asy(K,n,X,Tau,M$theta.hat,0.95,"Weibull")
AppCI<-Asy.phi(x0,t0,M$theta.hat,AsyCI$V,0.95,"Weibull")
```

$x_2 = \ln(\text{Volt})$, are commonly used for temperature (K) and voltage (V) in reliability studies. Furthermore, the R codes for the point estimation as well as for the 95% confidence intervals of all parameters of interest for the exponential, gamma, and Weibull lifetime distributions are displayed in Table 2.12. It needs to be mentioned that the R codes for the EM algorithm require the installation of package "expint" for exponential integral and function "integrate" for computation of $H_1(a, b)$ and $H_2(a, b)$. Finally, the outputs for all the considered distributions are presented in Tables 2.13 and 2.14. From these results, we can observe that the estimates of mean lifetime and reliability at three mission times are indeed different for the exponential, gamma, and Weibull lifetime distributions. In order to decide which one we should reply on, model selection and model mis-specification analysis discussed

Table 2.13 R outputs for inference based on electro-explosive devices data in Table 1.1 for exponential, gamma, and Weibull lifetime distributions.

	MLE	asymptotic	arsech	log/logit
e_0	11.739[a]	(−0.085, 23.563)[d]	—	—
e_1	−4746[a]	(−8519,−973)[d]	—	—
μ_E	62.816[b]	(7.142, 124)[e]	—	(26.987, 161)[f]
$R_E(10)$	0.859[c]	(0.743, 0.975)[e]	(0.733, 0.956)[f]	(0.700, 0.941)[f]
$R_E(20)$	0.738[c]	(0.538, 0.938)[e]	(0.543, 0.919)[f]	(0.500, 0.888)[f]
$R_E(30)$	0.634[c]	(0.376, 0.892)[e]	(0.405, 0.885)[f]	(0.363, 0.840)[f]
a_0	6.457[a]	(−28.001, 40.914)[d]	—	—
a_1	−1966[a]	(−12 936, 9003)[d]	—	—
b_0	−18.100[a]	(−62.979, 26.780)[d]	—	—
b_1	6669[a]	(−7755, 21 094)[d]	—	—
μ_G	62.725[b]	(0, 151)[e]	—	(15.405, 255)[f]
$R_G(10)$	0.823[c]	(0.387, 1.000)[e]	(0.396, 1.000)[f]	(0.190, 0.989)[f]
$R_G(20)$	0.696[c]	(0.301, 1.000)[e]	(0.356, 0.993)[f]	(0.262, 0.936)[f]
$R_G(30)$	0.593[c]	(0.266, 0.920)[e]	(0.322, 0.915)[f]	(0.273, 0.849)[f]
r_0	1.437[a]	(−20.077, 22.951)[d]	—	—
r_1	−399[a]	(−7278, 6479)[d]	—	—
s_0	−9.785[a]	(−22.661, 3.092)[d]	—	—
s_1	4118[a]	(−22.275, 8258)[d]	—	—
μ_W	54.469[b]	(0.000, 134)[e]	—	(12.581, 236)[f]
$R_W(10)$	0.862[c]	(0.520, 1.000)[e]	(0.489, 1.000)[f]	(0.260, 0.991)[f]
$R_W(20)$	0.727[c]	(0.373, 1.000)[e]	(0.406, 0.991)[f]	(0.309, 0.941)[f]
$R_W(30)$	0.608[c]	(0.295, 0.921)[e]	(0.343, 0.913)[f]	(0.294, 0.852)[f]

a) M$theta.hat.
b) M$ET0.hat.
c) M$R0.hat.
d) AsyCI$AsyCI.
e) AppCI$AsyCI.phi.
f) AppCI$AppCI.phi.

Table 2.14 R outputs for inference based on glass capacitors data in Table 1.2 for exponential, gamma, and Weibull lifetime distributions.

	MLE	asymptotic	arsech	log/logit
e_0	−23.471[a)]	(−73.884, 26.942)[d)]	—	—
e_1	−2669[a)]	(−23 801, 18 463)[d)]	—	—
e_2	3.916[a)]	(0.782, 7.049)[d)]	—	—
μ_E	2633[b)]	(113, 5154)[e)]	—	(1011, 6857)[f)]
$R_E(500)$	0.827[c)]	(0.677, 0.977)[e)]	(0.668, 0.953)[f)]	(0.626, 0.932)[f)]
$R_E(800)$	0.738[c)]	(0.523, 0.953)[e)]	(0.529, 0.929)[f)]	(0.482, 0.895)[f)]
$R_E(1000)$	0.684[c)]	(0.435, 0.933)[e)]	(0.454, 0.914)[f)]	(0.407, 0.872)[f)]
a_0	5.021[a)]	(−137, 147)[d)]	—	—
a_1	−1746[a)]	(−48 451, 44 960)[d)]	—	—
a_2	0.220[a)]	(−15.666, 16.106)[d)]	—	—
b_0	1.130[a)]	(−166, 168)[d)]	—	—
b_1	4730[a)]	(−48 744, 58 204)[d)]	—	—
b_2	−1.390[a)]	(−20.512, 17.731)[d)]	—	—
μ_G	617[b)]	(177, 1056)[e)]	—	(303, 1257)[f)]
$R_G(500)$	0.702[c)]	(0.197, 1.000)[e)]	(0.292, 1.000)[f)]	(0.174, 0.964)[f)]
$R_G(800)$	0.169[c)]	(0.000, 0.977)[e)]	(0.001, 1.000)[f)]	(0.001, 0.985)[f)]
$R_G(1000)$	0.040[c)]	(0.000, 0.404)[e)]	(0.000, 1.000)[f)]	(0.000, 0.998)[f)]
r_0	3.888[a)]	(−80.332, 88.108)[d)]	—	—
r_1	−1653[a)]	(−29 869, 26 562)[d)]	—	—
r_2	0.240[a)]	(−8.911, 9.390)[d)]	—	—
s_0	6.318[a)]	(−19.734, 32.370)[d)]	—	—
s_1	3108[a)]	(−4402, 10 619)[d)]	—	—
s_2	−1.243[a)]	(−4.513, 2.027)[d)]	—	—
μ_W	589[b)]	(255, 924)[e)]	—	(334, 1039)[f)]
$R_W(500)$	0.724[c)]	(0.264, 1.000)[e)]	(0.332, 1.000)[f)]	(0.208, 0.963)[f)]
$R_W(800)$	0.079[c)]	(0.000, 0.806)[e)]	(0.000, 1.000)[f)]	(0.000, 0.999)[f)]
$R_W(1000)$	0.001[c)]	(0.000, 0.038)[e)]	(0.000, 1.000)[f)]	(0.000, 1.000)[f)]

a) M$theta.hat.
b) M$ET0.hat.
c) M$R0.hat.
d) AsyCI$AsyCI.
e) AppCI$AsyCI.phi.
f) AppCI$AppCI.phi.

later in Chapter 4 will prove useful. In addition, from Table 2.14, we observe that the confidence intervals for reliability may not be informative and useful when the sample size is small. In this regard, by incorporating prior information, Bayesian approach discussed in Chapter 3 may provide more useful inference on mean lifetime and reliability, especially in the case of small sample sizes.

3

Bayesian Inference

3.1 Brief Overview

Bayesian inference, incorporating prior information, will be quite useful in developing inference for one-shot device testing data, especially in the case of small sample sizes. In the Bayesian setup, combination of extensive experience and expert opinions can provide prior information to form a framework for developing efficient inferential methods. The computational and methodological advantages make Bayesian approach a promising one for analyzing one-shot device testing data. This chapter describes the Bayesian inferential method for one-shot device testing data under constant-stress accelerated life-tests (CSALTs), as originally developed by Fan et al. (2009).

3.2 Bayesian Framework

Under the Bayesian framework (Fan et al., 2009), we start with the conditional joint likelihood function, given the data, $\mathbf{z}$,

$$L(\boldsymbol{\theta}|\mathbf{z}) \propto \prod_{i=1}^{I}(F(\tau_i;\boldsymbol{\theta}))^{n_i}(1 - F(\tau_i;\boldsymbol{\theta}))^{K_i - n_i}, \tag{3.1}$$

and assume we have $\pi(\boldsymbol{\theta})$ to be the joint prior density of $\boldsymbol{\theta}$. Then, the joint posterior distribution is given by

$$\pi(\boldsymbol{\theta}|\mathbf{z}) = \frac{q(\boldsymbol{\theta})}{\int_{\boldsymbol{\theta}\in\boldsymbol{\Theta}} q(\boldsymbol{\theta})d\boldsymbol{\theta}}, \tag{3.2}$$

where $q(\boldsymbol{\theta}) = L(\boldsymbol{\theta}|\mathbf{z})\pi(\boldsymbol{\theta})$ and $\boldsymbol{\Theta}$ is the parameter space of $\boldsymbol{\theta}$.

Accelerated Life Testing of One-shot Devices: Data Collection and Analysis, First Edition.
Narayanaswamy Balakrishnan, Man Ho Ling, and Hon Yiu So.

Companion Website: www.wiley.com/go/Balakrishnan/Accelerated_Life_Testing

Since the denominator is usually not in a closed-form, the posterior distribution and the subsequent Bayesian inference cannot be developed analytically in many situations. For this reason, Markov Chain Monte Carlo (MCMC) sampling scheme, using Metropolis–Hastings algorithm (Hastings, 1970), is often applied to generate posterior samples of $\boldsymbol{\theta}$, and then to approximate the posterior distribution and use it to develop the required Bayesian inference. The Metropolis–Hastings algorithm proceeds as follows: In the mth step of the iterative procedure,

Step 1: Generate $\boldsymbol{\theta}^*$ from a proposal distribution based on $\boldsymbol{\theta}^{(m)}$;
Step 2: Compute the acceptance probability $p = \min(1, q(\boldsymbol{\theta}^*)/q(\boldsymbol{\theta}^{(m)}))$;
Step 3: Set $\boldsymbol{\theta}^{(m+1)} = \begin{cases} \boldsymbol{\theta}^*, & \text{with probability p,} \\ \boldsymbol{\theta}^{(m)}, & \text{with probability } 1 - \text{p.} \end{cases}$

Suppose $M \geq 100\,000$ iterated samples of $\boldsymbol{\theta}$ are obtained from the posterior distribution as $\boldsymbol{\theta}^{(m)}$, for $m = 1, 2, \ldots, M$, using the above process. We then usually discard the first $D = 1000$ (say) samples as burn-in and sample one value in every $R = 100$ (say) iterations after the burn-in to reduce autocorrelation between the iterated samples in the MCMC sample. This way, we will end up with $B = \lfloor (M - D)/R \rfloor$ samples from the posterior distribution, where $\lfloor a \rfloor$ denotes the floor function of a, which is considered to be optimal in practice as stated by Roberts et al. (1997) and Roberts and Rosenthal (2001). Let $\boldsymbol{\theta}^{[b]}$, $b = 1, 2, \ldots, B$, be the posterior samples so obtained from the posterior distribution. The marginal sample means can then be used as approximate Bayesian point estimates of the parameters of interest, that is,

$$\hat{\boldsymbol{\theta}} = \frac{1}{B}\sum_{b=1}^{B} \boldsymbol{\theta}^{[b]}, \tag{3.3}$$

$$\hat{\mu}(\boldsymbol{\theta}) = \frac{1}{B}\sum_{b=1}^{B} \mu(\boldsymbol{\theta}^{[b]}), \tag{3.4}$$

$$\hat{R}(t; \boldsymbol{\theta}) = \frac{1}{B}\sum_{b=1}^{B} R(t; \boldsymbol{\theta}^{[b]}). \tag{3.5}$$

Furthermore, by sorting the posterior samples of $\{\phi(\boldsymbol{\theta}^{[b]}), b = 1, 2, \ldots, B\}$ to obtain $\phi^{[1]} \leq \phi^{[2]} \leq \cdots \leq \phi^{[B]}$, a $100(1 - \delta)\%$ credible interval for a parameter of interest $\phi(\boldsymbol{\theta})$ can be provided as

$$(\phi^{[\lfloor B\delta/2 \rfloor]},\ \phi^{[\lfloor B(1-\delta/2) \rfloor]}). \tag{3.6}$$

In the above, a credible interval is an interval in which an unobserved parameter value falls with a specified subjective probability.

3.3 Choice of Priors

In Bayesian inference, the prior information plays a very important role, especially when the data provide insufficient information about the parameters. Some commonly used priors and the corresponding hyperparameters, based on experts' information, are detailed below.

3.3.1 Laplace Prior

Fan et al. [2009] considered exponential prior distributions for the case of exponential lifetime distribution as the model parameters are known to be positive. But, in a general setting, the model parameters can be either positive or negative, and so we may naturally consider Laplace prior distributions on the model parameters for simplicity and convenience. Suppose there are G model parameters in the assumed model. Then, assuming independence between these model parameters, $\boldsymbol{\theta} = (\theta_1, \theta_2, \ldots, \theta_G)$, the joint prior density can be provided as

$$\pi_L(\boldsymbol{\theta}) \propto \exp\left\{-\sum_{g=1}^{G}\left(\left|\frac{\theta_g}{\theta_g^h}\right|\right)\right\}, \tag{3.7}$$

where θ_g^h's in (3.7) are the associated hyperparameters with $E[\theta_g] = \theta_g^h$, for $g = 1, 2, \ldots, G$. Consequently, the joint posterior distribution in (3.2) is given by

$$\begin{aligned}\pi_L(\boldsymbol{\theta}|\mathbf{z}) &\propto L(\boldsymbol{\theta}|\mathbf{z})\pi_L(\boldsymbol{\theta}) \\ &= (\prod_{i=1}^{I}(F(\tau_i;\boldsymbol{\theta}))^{n_i}(1-F(\tau_i;\boldsymbol{\theta}))^{K_i-n_i})\exp\left\{-\sum_{g=1}^{G}\left(\left|\frac{\theta_g}{\theta_g^h}\right|\right)\right\}. \end{aligned} \tag{3.8}$$

3.3.2 Normal Prior

As differences would naturally exist between the prior belief of the experts and the true values of the unknown parameters, we may assume

$$R_i^h = R(\tau_i;\boldsymbol{\theta}) + \epsilon_i,$$

where ϵ_i's are independently and identically distributed as normal with mean zero and variance σ^2, i.e. $N(0, \sigma^2)$. Then, the conditional likelihood function, given σ^2, is

$$L(\boldsymbol{\theta};\mathbf{z}, R_i^h, \sigma^2) \propto \prod_{i=1}^{I}\frac{1}{\sqrt{2\pi\sigma^2}}\exp\left(-\frac{(R_i^h - R(\tau_i;\boldsymbol{\theta}))^2}{2\sigma^2}\right).$$

We can now adopt this as the prior of $\boldsymbol{\theta}$, on the basis of the available reliability R_i^h, that is,

$$\pi_N(\boldsymbol{\theta}|\mathbf{z}, R_i^h, \sigma^2) \propto \prod_{i=1}^{I} \frac{1}{\sqrt{2\pi\sigma^2}} \exp\left(-\frac{(R_i^h - R(\tau_i;\boldsymbol{\theta}))^2}{2\sigma^2}\right). \tag{3.9}$$

In (3.9), we have a data-based prior (Zellner, 1971) and it takes into account the form of the experiment used to generate the observational data. Since σ^2 is unknown, we can use, for example the non-informative prior

$$\pi(\sigma^2) \propto \frac{1}{\sigma^2}, \quad \sigma^2 > 0,$$

resulting in the joint prior density as

$$\begin{aligned} \pi_N(\boldsymbol{\theta}|\mathbf{z}, R_i^h) &\propto \int_0^\infty \pi_N(\boldsymbol{\theta}|\mathbf{z}, R_i^h, \sigma^2)\pi(\sigma^2)d\sigma^2 \\ &\propto \int_0^\infty (\sigma^2)^{-\left(\frac{I}{2}+1\right)} \exp\left(-\frac{\sum_{i=1}^{I}(R_i^h - R(\tau_i;\boldsymbol{\theta}))^2}{2\sigma^2}\right) d\sigma^2 \\ &\propto \left(\sum_{i=1}^{I}(R_i^h - R(\tau_i;\boldsymbol{\theta}))^2\right)^{-I/2}. \end{aligned} \tag{3.10}$$

The joint posterior distribution in (3.2) then becomes

$$\begin{aligned} \pi_N(\boldsymbol{\theta}|\mathbf{z}) &\propto L(\boldsymbol{\theta}|\mathbf{z})\pi_N(\boldsymbol{\theta}) \\ &= \left(\prod_{i=0}^{I}(F(\tau_i;\boldsymbol{\theta}))^{n_i}(1 - F(\tau_i;\boldsymbol{\theta}))^{K_i-n_i}\right)\left(\sum_{i=1}^{I}(R_i^h - R(\tau_i;\boldsymbol{\theta}))^2\right)^{-I/2}. \end{aligned}$$

3.3.3 Beta Prior

A conjugate prior, which yields the posterior distribution to be in the same family of the prior density, offers mathematical convenience in addition to yielding a closed-form expression for the posterior distribution. In the case of one-shot device testing data, Fan et al. (2009) and Mun et al. (2019) considered beta prior as the conjugate priors for $R(\tau_i;\boldsymbol{\theta})$ as follows:

$$\pi_B(\boldsymbol{\theta}) \propto \prod_{i=1}^{I}(R(\tau_i;\boldsymbol{\theta}))^{c_i-1}(1 - R(\tau_i;\boldsymbol{\theta}))^{d_i-1}, \tag{3.11}$$

where the hyperparameters c_i and d_i may be chosen so that

$$E[R(\tau_i;\boldsymbol{\theta})] = \frac{c_i}{c_i + d_i} = R_i^h,$$

$$V[R(\tau_i;\boldsymbol{\theta})] = \frac{c_i d_i}{(c_i + d_i)^2(c_i + d_i + 1)} = \sigma_R^2,$$

with σ_R^2 representing the uncertainty in the experts' belief. So, given R_i^h and σ_R^2, the hyperparameters then become

$$c_i = R_i^h \left(\frac{R_i^h(1 - R_i^h)}{\sigma_R^2} - 1 \right),$$

$$d_i = (1 - R_i^h) \left(\frac{R_i^h(1 - R_i^h)}{\sigma_R^2} - 1 \right).$$

Finally, the posterior distribution in (3.2), upon incorporating the prior in (3.11), becomes

$$\begin{aligned} \pi_B(\boldsymbol{\theta}|\mathbf{z}) &\propto L(\boldsymbol{\theta}|\mathbf{z})\pi_B(\boldsymbol{\theta}) \\ &= \prod_{i=0}^{I} (F(\tau_i; \boldsymbol{\theta}))^{n_i+d_i-1}(1 - F(\tau_i; \boldsymbol{\theta}))^{K_i-n_i+c_i-1}. \end{aligned} \tag{3.12}$$

It is worth noting that the beta prior may be viewed as providing $c_i + d_i - 2 = R_i^h(1 - R_i^h)/\sigma_R^2 - 3$ additional devices in accelerated life-tests (ALTs), and so there is a restriction on the uncertainty of prior belief that $\sigma_R^2 < 3R_i^h(1 - R_i^h)$. Moreover, the beta prior is expected to dominate the posterior information when the uncertainty in prior belief, σ_R^2, is small.

Finally, all of Laplace, normal and beta priors described above are suitable for the analysis of one-shot device testing data for exponential, gamma, and Weibull lifetime distributions as demonstrated in the following sections.

3.4 Simulation Studies

In this section, extensive simulation studies are performed to evaluate the performance of the proposed Bayesian estimation methods for one-shot devices under CSALTs listed in Table 2.3, for various sample sizes under different lifetime distributions, in terms of bias, root mean square errors (RMSEs), coverage probabilities, and average widths of 95% credible intervals, based on 1000 Monte Carlo simulations.

In the Metropolis–Hastings algorithm, the number of iterated samples, the number of discarded samples as burn-in, and the number of iterations for sampling one value after the burn-in were set to be $M = 100\,000, D = 1000$, and $R = 100$, respectively. In addition, the normal distribution with mean $\boldsymbol{\theta}^{(m)}$ and standard deviation $1 \times 10^{c-1}$, where c is the position of the first nonzero digit of $\boldsymbol{\theta}^{(0)}$, was adopted as the proposal distribution for the parameters to generate $\boldsymbol{\theta}^*$ in Step 1. For example if $\theta^{(0)} = 0.0712$, the corresponding standard deviation is 0.001 and θ^* is then generated from $N(0.0712, 0.001^2)$. The values of the hyperparameters

Table 3.1 The values of the hyperparameters for Laplace, normal, and beta priors for exponential, gamma, and Weibull lifetime distributions.

Exponential lifetime distribution with $\boldsymbol{\theta}_E = (e_0, e_1, e_2) = (-6, 0.03, 0.03)$	
$\theta_L(\boldsymbol{\theta}_E)$	$(e_0^h, e_1^h, e_2^h) = (-5.8, 0.025, 0.025)$
$\theta_N(\boldsymbol{\theta}_E)/\theta_B(\boldsymbol{\theta}_E)$	$(R_1^h, R_2^h, R_3^h, R_4^h, R_5^h, R_6^h, R_7^h, R_8^h, R_9^h, R_{10}^h, R_{11}^h, R_{12}^h) = (0.8, 0.8, 0.8, 0.8, 0.7, 0.6, 0.6, 0.6, 0.6, 0.5, 0.5, 0.4)$
Gamma lifetime distribution with $\boldsymbol{\theta}_G = (a_0, a_1, a_2, b_0, b_1, b_2) = (-0.5, 0.02, 0.02, 7, -0.06, -0.06)$	
$\theta_L(\boldsymbol{\theta}_G)$	$(a_0^h, a_1^h, a_2^h, b_0^h, b_1^h, b_2^h) = (-0.7, 0.025, 0.025, 7.2, -0.065, -0.065)$
$\theta_N(\boldsymbol{\theta}_G)/\theta_B(\boldsymbol{\theta}_G)$	$(R_1^h, R_2^h, R_3^h, R_4^h, R_5^h, R_6^h, R_7^h, R_8^h, R_9^h, R_{10}^h, R_{11}^h, R_{12}^h) = (0.9, 0.9, 0.9, 0.8, 0.8, 0.6, 0.6, 0.4, 0.6, 0.4, 0.4, 0.1)$
Weibull lifetime distribution with $\boldsymbol{\theta}_W = (r_0, r_1, r_2, s_0, s_1, s_2) = (-1, 0.02, 0.02, 6, -0.03, -0.03)$	
$\theta_L(\boldsymbol{\theta}_W)$	$(r_0^h, r_1^h, r_2^h, s_0^h, s_1^h, s_2^h) = (-1.5, 0.025, 0.025, 5.5, -0.025, -0.025)$
$\theta_N(\boldsymbol{\theta}_W)/\theta_B(\boldsymbol{\theta}_W)$	$(R_1^h, R_2^h, R_3^h, R_4^h, R_5^h, R_6^h, R_7^h, R_8^h, R_9^h, R_{10}^h, R_{11}^h, R_{12}^h) = (0.9, 0.9, 0.9, 0.9, 0.7, 0.6, 0.6, 0.6, 0.5, 0.4, 0.4, 0.3)$

for Laplace, normal, and beta priors for exponential, gamma, and Weibull lifetime distributions are all listed in Table 3.1. Moreover, the uncertainty in beta prior σ_R^2 was taken to be 0.01.

Tables 3.2–3.4 present the bias and RMSEs of the Bayesian estimates of the model parameters, mean lifetime, and reliability under normal operating conditions under the exponential, gamma, and Weibull lifetime distributions with the three priors described in the preceding section. From these tables, we observe that the normal prior performs well for the estimation of the parameters of interest for the exponential and gamma distributions. But, the Laplace prior outperforms the normal and beta priors for the estimation of mean lifetime and reliability under the Weibull distribution.

Tables 3.5-3.7 present the coverage probabilities and average widths of the credible intervals for the model parameters, mean lifetime, and reliability under normal operating conditions for the exponential, gamma, and Weibull lifetime

Table 3.2 Bias and RMSEs of Bayesian estimates of the model parameter, θ_E, mean lifetime, μ_E, and reliability, $R_E(t)$, under normal operating conditions $\mathbf{x}_0 = (25, 35)$, for various sample sizes, K, under exponential lifetime distribution.

		Parameter	e_0	e_1	e_2	μ_E	$R_E(10)$	$R_E(20)$	$R_E(30)$
K	**Prior**	**True value**	**−6.000**	**0.030**	**0.030**	**66.686**	**0.799**	**0.638**	**0.509**
5	Laplace	Bias	1.774	−0.022	−0.023	−16.076	−0.108	−0.146	−0.150
		RMSE	2.007	0.028	0.028	24.276	0.130	0.174	0.180
	Normal	Bias	0.582	−0.009	−0.009	−0.755	−0.004	−0.006	−0.007
		RMSE	0.586	0.009	0.009	1.497	0.005	0.008	0.009
	Beta	Bias	0.198	−0.005	−0.005	11.038	0.008	0.017	0.024
		RMSE	0.598	0.011	0.011	16.297	0.027	0.045	0.055
10	Laplace	Bias	1.410	−0.018	−0.018	−14.527	−0.084	−0.118	−0.124
		RMSE	1.680	0.023	0.024	20.821	0.103	0.144	0.153
	Normal	Bias	0.572	−0.009	−0.009	−0.818	−0.004	−0.006	−0.007
		RMSE	0.579	0.009	0.009	1.778	0.006	0.009	0.011
	Beta	Bias	0.181	−0.004	−0.004	8.949	0.006	0.012	0.018
		RMSE	0.650	0.011	0.012	15.619	0.029	0.047	0.057
30	Laplace	Bias	0.751	−0.009	−0.010	−7.905	−0.043	−0.062	−0.068
		RMSE	1.121	0.017	0.017	16.468	0.062	0.092	0.103
	Normal	Bias	0.515	−0.008	−0.008	−0.593	−0.004	−0.005	−0.006
		RMSE	0.539	0.009	0.009	2.958	0.008	0.013	0.015
	Beta	Bias	0.088	−0.002	−0.002	6.149	0.004	0.008	0.013
		RMSE	0.662	0.011	0.011	14.542	0.031	0.049	0.060

Table 3.3 Bias and RMSEs of Bayesian estimates of the model parameter, $\boldsymbol{\theta}_G$, mean lifetime, μ_G, and reliability, $R_G(t)$, under normal operating conditions $\mathbf{x}_0 = (25, 35)$, for various sample sizes, K, under gamma lifetime distribution.

		Parameter	a_0	a_1	a_2	b_0	b_1	b_2	μ_G	$R_G(10)$	$R_G(20)$	$R_G(30)$
K	**Prior**	**True value**	**−0.500**	**0.020**	**0.020**	**7.000**	**−0.060**	**−0.060**	**60.340**	**0.912**	**0.739**	**0.562**
30	Laplace	Bias	0.724	−0.010	−0.010	−0.826	0.012	0.011	1.801	0.007	0.002	−0.010
		RMSE	0.861	0.016	0.015	1.180	0.022	0.019	12.750	0.028	0.060	0.086
	Normal	Bias	−0.187	0.002	0.000	0.394	−0.005	−0.002	5.650	−0.010	0.002	0.019
		RMSE	0.554	0.010	0.010	0.759	0.013	0.012	6.437	0.021	0.017	0.023
	Beta	Bias	−0.023	0.002	0.000	0.218	−0.004	−0.002	5.843	0.005	0.018	0.023
		RMSE	0.879	0.027	0.024	1.166	0.032	0.028	11.147	0.030	0.046	0.057
50	Laplace	Bias	0.644	−0.009	−0.008	−0.761	0.010	0.010	0.110	0.009	0.004	−0.009
		RMSE	0.783	0.015	0.013	1.077	0.020	0.018	9.918	0.024	0.050	0.072
	Normal	Bias	−0.184	0.002	0.001	0.372	−0.004	−0.002	5.107	−0.010	0.001	0.017
		RMSE	0.538	0.010	0.010	0.727	0.013	0.012	5.965	0.020	0.018	0.022
	Beta	Bias	−0.038	0.001	0.000	0.174	−0.003	−0.001	4.240	0.001	0.009	0.013
		RMSE	0.831	0.023	0.020	1.100	0.028	0.024	9.936	0.030	0.042	0.052
100	Laplace	Bias	0.514	−0.008	−0.006	−0.588	0.009	0.006	−0.402	0.011	0.010	−0.001
		RMSE	0.676	0.014	0.012	0.863	0.018	0.015	6.604	0.021	0.036	0.050
	Normal	Bias	−0.145	0.001	0.001	0.305	−0.003	−0.002	4.256	−0.007	0.002	0.015
		RMSE	0.507	0.010	0.009	0.677	0.012	0.011	5.325	0.018	0.017	0.022
	Beta	Bias	0.003	0.000	0.001	0.091	0.000	−0.002	2.596	0.002	0.008	0.010
		RMSE	0.695	0.018	0.016	0.910	0.021	0.019	7.277	0.024	0.033	0.041

Table 3.4 Bias and RMSEs of Bayesian estimates of the model parameter, θ_W, mean lifetime, μ_W, and reliability, $R_W(t)$, under normal operating conditions $\mathbf{x}_0 = (25, 35)$, for various sample sizes, K, under Weibull lifetime distribution.

		Parameter	r_0	r_1	r_2	s_0	s_1	s_2	μ_W	$R_W(10)$	$R_W(20)$	$R_W(30)$
K	**Prior**	**True value**	**−1.000**	**0.020**	**0.020**	**6.000**	**−0.030**	**−0.030**	**6.245e1**	**0.851**	**0.686**	**0.539**
30	Laplace	Bias	1.205	−0.015	−0.016	−0.823	0.010	0.011	−8.035	0.021	−0.022	−0.083
		RMSE	1.302	0.019	0.019	1.027	0.015	0.015	18.177	0.045	0.073	0.126
	Normal	Bias	0.642	−0.008	−0.008	−1.098	0.012	0.012	6.778e6	−0.035	−0.108	−0.169
		RMSE	0.835	0.012	0.013	1.173	0.014	0.014	2.134e8	0.043	0.112	0.174
	Beta	Bias	0.387	−0.004	−0.005	−0.517	0.005	0.006	4.397e5	−0.013	−0.050	−0.103
		RMSE	1.335	0.024	0.023	0.876	0.014	0.013	1.389e7	0.061	0.071	0.129
50	Laplace	Bias	1.033	−0.012	−0.013	−0.641	0.008	0.008	−8.196	0.025	−0.008	−0.063
		RMSE	1.152	0.017	0.017	0.840	0.012	0.012	13.673	0.040	0.055	0.101
	Normal	Bias	0.555	−0.007	−0.007	−0.934	0.010	0.011	−1.386	−0.028	−0.089	−0.143
		RMSE	0.797	0.012	0.013	1.027	0.012	0.012	3.294e2	0.039	0.095	0.152
	Beta	Bias	0.328	−0.003	−0.004	−0.408	0.004	0.005	4.069e3	−0.007	−0.034	−0.076
		RMSE	1.105	0.019	0.020	0.743	0.011	0.011	1.285e5	0.049	0.055	0.102
100	Laplace	Bias	0.785	−0.009	−0.010	−0.429	0.005	0.005	−6.345	0.025	0.004	−0.037
		RMSE	0.936	0.014	0.014	0.611	0.009	0.009	11.499	0.037	0.039	0.068
	Normal	Bias	0.378	−0.005	−0.005	−0.587	0.007	0.007	−7.925	−0.014	−0.051	−0.089
		RMSE	0.755	0.012	0.012	0.749	0.010	0.009	16.856	0.034	0.063	0.104
	Beta	Bias	0.206	−0.003	−0.002	−0.248	0.003	0.003	−0.215	−0.002	−0.017	−0.043
		RMSE	0.828	0.015	0.014	0.543	0.009	0.008	15.779	0.038	0.038	0.066

Table 3.5 Coverage probabilities (CP) and average widths (AW) of 95% credible intervals for the model parameter, θ_E, mean lifetime, μ_E, and reliability, $R_E(t)$, under normal operating conditions $\mathbf{x}_0 = (25, 35)$, for various sample sizes, K, for exponential lifetime distribution.

		Parameter	e_0	e_1	e_2	μ_E	$R_E(10)$	$R_E(20)$	$R_E(30)$
K	**Prior**	**True value**	**−6.000**	**0.030**	**0.030**	**66.686**	**0.799**	**0.638**	**0.509**
5	Laplace	CP	0.824	0.901	0.907	0.870	0.870	0.870	0.870
		AW	5.723	0.101	0.100	90.988	0.383	0.501	0.514
	Normal	CP	0.972	0.988	0.991	1	1	1	1
		AW	1.571	0.027	0.027	28.070	0.077	0.122	0.146
	Beta	CP	1	0.997	0.999	1	1	1	1
		AW	4.493	0.077	0.078	96.580	0.207	0.325	0.387
10	Laplace	CP	0.826	0.911	0.886	0.861	0.861	0.861	0.861
		AW	4.817	0.084	0.084	74.697	0.301	0.415	0.440
	Normal	CP	0.957	0.978	0.974	1	1	1	1
		AW	1.556	0.027	0.027	27.795	0.077	0.121	0.144
	Beta	CP	0.996	0.995	0.994	0.998	0.998	0.998	0.998
		AW	4.057	0.07	0.07	84.685	0.189	0.298	0.354
30	Laplace	CP	0.854	0.895	0.896	0.862	0.862	0.862	0.862
		AW	3.391	0.059	0.058	56.205	0.188	0.280	0.315
	Normal	CP	0.903	0.919	0.936	1	1	1	1
		AW	1.539	0.027	0.026	27.511	0.075	0.119	0.141
	Beta	CP	0.976	0.976	0.979	0.976	0.976	0.976	0.976
		AW	3.073	0.053	0.053	61.418	0.145	0.229	0.273

Table 3.6 Coverage probabilities (CP) and average widths (AW) of 95% credible intervals for the model parameter, θ_G, mean lifetime, μ_G, and reliability, $R_G(t)$, under normal operating conditions $\mathbf{x}_0 = (25, 35)$, for various sample sizes, K, for gamma lifetime distribution.

		Parameter	a_0	a_1	a_2	b_0	b_1	b_2	μ_G	$R_G(10)$	$R_G(20)$	$R_G(30)$
K	**Prior**	**True value**	**−0.500**	**0.020**	**0.020**	**7.000**	**−0.060**	**−0.060**	**60.340**	**0.912**	**0.739**	**0.562**
30	Laplace	CP	0.725	0.954	0.937	0.874	0.951	0.954	0.959	0.944	0.958	0.957
		AW	1.914	0.068	0.056	3.629	0.088	0.076	53.362	0.112	0.246	0.349
	Normal	CP	0.821	0.982	0.963	0.839	0.977	0.969	0.957	0.924	1	1
		AW	1.369	0.050	0.042	2.026	0.060	0.051	22.208	0.065	0.098	0.125
	Beta	CP	0.798	0.893	0.865	0.885	0.896	0.893	0.971	0.919	0.961	0.977
		AW	2.06	0.086	0.071	3.384	0.104	0.088	45.165	0.101	0.188	0.262
50	Laplace	CP	0.688	0.925	0.935	0.841	0.940	0.930	0.941	0.922	0.943	0.935
		AW	1.739	0.061	0.051	2.999	0.077	0.067	38.056	0.090	0.193	0.275
	Normal	CP	0.818	0.975	0.961	0.847	0.974	0.958	0.942	0.911	0.997	0.998
		AW	1.337	0.047	0.040	1.958	0.058	0.049	20.526	0.062	0.093	0.118
	Beta	CP	0.803	0.891	0.893	0.877	0.905	0.903	0.957	0.904	0.944	0.959
		AW	1.925	0.075	0.063	3.027	0.091	0.078	36.570	0.091	0.162	0.222
100	Laplace	CP	0.711	0.922	0.920	0.833	0.908	0.938	0.943	0.900	0.938	0.950
		AW	1.514	0.051	0.044	2.375	0.063	0.055	25.907	0.068	0.136	0.195
	Normal	CP	0.818	0.970	0.942	0.846	0.968	0.957	0.932	0.908	0.984	0.991
		AW	1.236	0.042	0.036	1.783	0.051	0.044	17.654	0.055	0.083	0.105
	Beta	CP	0.799	0.910	0.898	0.863	0.909	0.902	0.947	0.895	0.938	0.960
		AW	1.643	0.059	0.050	2.462	0.072	0.061	26.100	0.073	0.126	0.171

Table 3.7 Coverage probabilities (CP) and average widths (AW) of 95% credible intervals for the model parameter, θ_W, mean lifetime, μ_W, and reliability, $R_W(t)$, under normal operating conditions $\mathbf{x}_0 = (25, 35)$, for various sample sizes, K, for Weibull lifetime distribution.

		Parameter	r_0	r_1	r_2	s_0	s_1	s_2	μ_W	$R_W(10)$	$R_W(20)$	$R_W(30)$
K	**Prior**	**True value**	**−1.000**	**0.020**	**0.020**	**6.000**	**−0.030**	**−0.030**	**6.245e1**	**0.851**	**0.686**	**0.539**
30	Laplace	CP	0.776	0.919	0.915	0.834	0.902	0.882	0.877	0.953	0.950	0.902
		AW	3.442	0.070	0.065	3.027	0.053	0.051	68.149	0.179	0.295	0.432
	Normal	CP	0.951	0.986	0.974	0.313	0.533	0.519	0.431	0.952	0.294	0.203
		AW	3.195	0.057	0.057	1.775	0.029	0.029	8.335e4	0.167	0.163	0.254
	Beta	CP	0.923	0.937	0.935	0.878	0.931	0.919	0.892	0.944	0.901	0.839
		AW	4.995	0.091	0.089	3.056	0.052	0.05	3.645e5	0.230	0.249	0.374
50	Laplace	CP	0.779	0.911	0.906	0.829	0.897	0.892	0.864	0.933	0.959	0.912
		AW	3.08	0.061	0.058	2.496	0.043	0.041	52.897	0.150	0.226	0.343
	Normal	CP	0.950	0.966	0.967	0.463	0.667	0.648	0.550	0.949	0.455	0.341
		AW	3.111	0.056	0.055	1.878	0.030	0.029	110.841	0.159	0.162	0.252
	Beta	CP	0.942	0.940	0.939	0.888	0.914	0.916	0.897	0.956	0.917	0.866
		AW	4.191	0.075	0.075	2.626	0.043	0.043	5.092e3	0.196	0.200	0.304
100	Laplace	CP	0.788	0.899	0.891	0.851	0.911	0.908	0.860	0.902	0.947	0.916
		AW	2.542	0.049	0.047	1.903	0.032	0.031	41.612	0.119	0.156	0.237
	Normal	CP	0.940	0.955	0.958	0.702	0.822	0.830	0.748	0.950	0.712	0.615
		AW	2.792	0.050	0.050	1.840	0.029	0.029	50.301	0.138	0.146	0.220
	Beta	CP	0.931	0.936	0.947	0.903	0.914	0.934	0.912	0.941	0.929	0.900
		AW	3.073	0.056	0.056	1.995	0.033	0.033	61.644	0.147	0.144	0.210

distributions with the three priors described in the preceding section. From these tables, we observe that the normal prior performs well with the credible interval for the parameters of interest for the exponential and gamma distributions. The estimation of mean lifetime and reliability, with the use of the Laplace prior, is satisfactory for the Weibull distribution. It is worth noting that, unlike the confidence intervals discussed in Section 2.4, credible intervals may not maintain the nominal level even in case of large sample sizes. In fact, we treat an unknown parameter as a fixed value in a confidence interval, indicating that the interval contains the (fixed, but unknown) parameter with a level of confidence, while the unknown parameter is treated as a random variable in a credible interval, indicating there is a chance of having the parameter within the interval. But, we do observe that the average widths of credible intervals decrease with increasing sample size for all lifetime distributions and all priors considered.

It is important to note, however, that the Bayesian estimates are sensitive to the choice of priors and the hyperparameters, especially when the sample size in the CSALT is small. This is so because, as mentioned earlier, the prior information and belief would dominate the ensuing Bayesian inference. Moreover, we observe that the choice of the standard deviation of the proposal distribution specified in Step 1 of the Metropolis–Hastings algorithm and the initial point, $\theta^{(0)}$, in the algorithm have a significant impact on the quality of approximation of the posterior distribution. A large M is usually specified for the Metropolis–Hastings algorithm to improve the accuracy of the approximation.

3.5 Case Study with R Codes

In this section, we consider the glass capacitors data under CSALTs presented in Table 1.2. The values of the hyperparameters chosen for Laplace, normal, and beta priors for exponential, gamma, and Weibull lifetime distributions are all listed in Table 3.8. In the Metropolis–Hastings algorithm, we set $M = 1\,000\,000, D = 10\,000$, and $R = 1000$. In addition, the normal distribution, $N(\theta^{(m)}, 0.001^2)$, was

Table 3.8 The values of the hyperparameters for Laplace, normal, and beta priors for exponential, gamma, and Weibull lifetime distributions for glass capacitors data presented in Table 1.2.

Laplace	$(e_0^h, e_1^h, e_2^h) = (-23.5, -2700, 4)$
	$(a_0^h, a_1^h, a_2^h, b_0^h, b_1^h, b_2^h) = (5, -1750, 0.22, 1.1, 4750, -1.4)$
	$(r_0^h, r_1^h, r_2^h, s_0^h, s_1^h, s_2^h) = (4, -1650, 0.24, 6.3, 3100, -1.25)$
Normal/Beta	$(R_1^h, R_2^h, R_3^h, R_4^h, R_5^h, R_6^h, R_7^h, R_8^h)$ $= (0.9, 0.9, 0.9, 0.9, 0.6, 0.6, 0.6, 0.6)$

Table 3.9 R codes for the estimation of parameters and 95% credible intervals, with different priors, for the model parameters, mean lifetime, and reliability at mission times under normal operating conditions, for exponential, gamma, and Weibull lifetime distributions, for the glass capacitors data in Table 1.2.

```
source(”Baye.R”)

#Metropolis-Hastings algorithm
M<-1e6; D<-1e4; R<-1e3

#hyperparameters
h.EL<-c(-23.5,-2700,4)
h.GL<-c(5,-1750,0.22,1.1,4750,-1.4)
h.WL<-c(4,-1650,0.24,6.3,3100,-1.25)
h.N<-c(0.9,0.9,0.9,0.9,0.6,0.6,0.6,0.6)
h.B<-c(h.N,0.01)

#exponential distribution with Laplace prior
sdp<-rep(1e-3,3) #sd in proposal distribution
BL<-Baye(K,n,X,Tau,h.EL,x0,t0,M,loc=0.95,”exp”,”Lap”,h.EL,D,R,sdp)
#exponential distribution with normal prior
BN<-Baye(K,n,X,Tau,h.EL,x0,t0,M,loc=0.95,”exp”,”norm”,h.N,D,R,sdp)
#exponential distribution with beta prior
BB<-Baye(K,n,X,Tau,h.EL,x0,t0,M,loc=0.95,”exp”,”beta”,h.B,D,R,sdp)

#gamma distribution with Laplace prior
sdp<-rep(1e-3,6) #sd in proposal distribution
BL<-Baye(K,n,X,Tau,h.GL,x0,t0,M,loc=0.95,”gamma”,”Lap”,h.GL,D,R,sdp)
#gamma distribution with normal prior
BN<-Baye(K,n,X,Tau,h.GL,x0,t0,M,loc=0.95,”gamma”,”norm”,h.N,D,R,sdp)
#gamma distribution with beta prior
BB<-Baye(K,n,X,Tau,h.GL,x0,t0,M,loc=0.95,”gamma”,”beta”,h.B,D,R,sdp)

#Weibull distribution with Laplace prior
BL<-Baye(K,n,X,Tau,h.WL,x0,t0,M,loc=0.95,”Weibull”,”Lap”,h.WL,D,R,sdp)
#Weibull distribution with normal prior
BN<-Baye(K,n,X,Tau,h.WL,x0,t0,M,loc=0.95,”Weibull”,”norm”,h.N,D,R,sdp)
#Weibull distribution with beta prior
BB<-Baye(K,n,X,Tau,h.WL,x0,t0,M,loc=0.95,”Weibull”,”beta”,h.B,D,R,sdp)
```

Table 3.10 Bayesian estimates for glass capacitors data in Table 1.2 for exponential, gamma, and Weibull lifetime distributions, obtained by using the R codes in Table 3.9.

Parameter	Laplace[a]	Normal[b]	Beta[c]
e_0	−23.250	−23.201	−22.447
e_1	−2700	−2699	−2699
e_2	3.872	3.905	3.754
μ_E	2993	2289	2490
$R_E(500)$	0.836	0.803	0.815
$R_E(800)$	0.752	0.704	0.722
$R_E(1000)$	0.701	0.645	0.665
a_0	4.356	5.575	5.663
a_1	−1750	−1750	−1750
a_2	0.221	0.017	0.118
b_0	0.741	1.318	1.148
b_1	4751	4749	4751
b_2	−1.182	−1.313	−1.406
μ_G	733	667	650
$R_G(500)$	0.688	0.689	0.747
$R_G(800)$	0.296	0.277	0.200
$R_G(1000)$	0.164	0.124	0.069
r_0	4.528	4.562	4.901
r_1	−1650	−1651	−1650
r_2	−0.092	0.048	0.063
s_0	5.838	6.765	5.881
s_1	3100	3100	3100
s_2	−0.973	−1.305	−1.160
μ_W	7374	628	595
$R_W(500)$	0.737	0.697	0.727
$R_W(800)$	0.483	0.224	0.088
$R_W(1000)$	0.401	0.060	0.021

a) BL$phi.hat.
b) BN$phi.hat.
c) BB$phi.hat.

Table 3.11 95% credible intervals for glass capacitors data in Table 1.2 for exponential, gamma, and Weibull lifetime distributions, obtained by using the R codes in Table 3.9.

Parameter	Laplace[a)]	Normal[b)]	Beta[c)]
e_0	(−23.637,−22.719)	(−24.007,−22.256)	(−23.227,−21.937)
e_1	(−2701,−2700)	(−2700,−2698)	(−2700,−2699)
e_2	(3.731, 4.007)	(3.730, 4.050)	(3.643, 3.887)
μ_E	(1701, 4974)	(1975, 2674)	(1927, 3264)
$R_E(500)$	(0.745, 0.904)	(0.776, 0.829)	(0.771, 0.858)
$R_E(800)$	(0.625, 0.851)	(0.667, 0.741)	(0.660, 0.783)
$R_E(1000)$	(0.555, 0.818)	(0.603, 0.688)	(0.595, 0.736)
a_0	(3.613, 5.008)	(4.878, 6.247)	(4.934, 6.290)
a_1	(−1751,−1749)	(−1751,−1750)	(−1751,−1749)
a_2	(−0.088, 0.383)	(−0.140, 0.214)	(−0.099, 0.371)
b_0	(0.405, 1.167)	(0.915, 1.799)	(0.402, 2.214)
b_1	(4750, 4751)	(4748, 4750)	(4750, 4751)
b_2	(−1.409,−0.867)	(−1.396,−1.204)	(−1.691,−1.000)
μ_G	(545, 1456)	(604, 726)	(569, 824)
$R_G(500)$	(0.557, 0.803)	(0.649, 0.718)	(0.658, 0.838)
$R_G(800)$	(0.055, 0.612)	(0.160, 0.358)	(0.027, 0.436)
$R_G(1000)$	(0.005, 0.530)	(0.040, 0.199)	(0.001, 0.293)
r_0	(4.033, 5.006)	(3.918, 5.026)	(3.931, 6.579)
r_1	(−1650,−1649)	(−1651,−1650)	(−1651,−1650)
r_2	(−0.342, 0.190)	(−0.042, 0.177)	(−0.241, 0.283)
s_0	(5.227, 6.266)	(6.245, 7.425)	(5.258, 6.487)
s_1	(3099, 3100)	(3099, 3100)	(3099, 3100)
s_2	(−1.218,−0.574)	(−1.420,−1.218)	(−1.277,−1.041)
μ_W	(546, 51962)	(589, 678)	(544, 727)
$R_W(500)$	(0.615, 0.837)	(0.662, 0.732)	(0.645, 0.804)
$R_W(1000)$	(0.007, 0.769)	(0.109, 0.330)	(0.000, 0.389)
$R_W(1500)$	(0.000, 0.738)	(0.005, 0.152)	(0.000, 0.217)

a) BL$BCI.
b) BN$BCI.
c) BB$BCI.

adopted to generate θ^* in Step 1 of the algorithm. Moreover, the uncertainty in beta prior σ_R^2 was taken to be 0.01.

The R codes for defining the glass capacitors data are displayed in Table 2.11. The R codes for the Bayesian estimation and 95% credible intervals for the model parameters, mean lifetime, and reliability at the mission times under normal operating conditions, for the exponential, gamma, and Weibull lifetime distributions with Laplace, normal, and beta priors, are presented in Table 3.9. Finally, the Bayesian estimates and credible intervals obtained for all these cases are given in Tables 3.10 and 3.11, respectively. Compared with the confidence intervals reported in Table 2.14, the credible intervals provide more useful information on the mean lifetime and reliability under gamma and Weibull distributions. But, the Bayesian estimate of a parameter may not lie within the credible interval, due to extreme values of iterated samples existing in the MCMC sample. For this reason, the Bayesian estimate may be obtained by the sample median of the MCMC sample, $\hat{\phi}(\theta) = \phi^{[\lfloor B/2 \rfloor]}$, rather than the sample mean of the MCMC sample. Moreover, we observe a wide range in the Bayesian estimates of the model parameters, mean lifetime and reliability for different priors and different lifetime distributions. Also, the corresponding 95% credible intervals are quite different, especially for the mean lifetime. The Bayesian estimates and the credible intervals are quite sensitive to the choice of lifetime distribution and prior for these data. Unlike in the case of the EM algorithm in Chapter 2, the Bayesian estimates and the credible intervals are non-reproducible, meaning that we cannot obtain the same results for another iterative sequence, especially when the number of parameters to be estimated is quite large.

4

Model Mis-Specification Analysis and Model Selection

4.1 Brief Overview

Model-based methods form a key part of reliability analysis. Two of the popular distributions used for modeling lifetime data are gamma and Weibull since both are flexible enough to model lifetimes with increasing, decreasing, and constant failure rates. Besides, both of them include the exponential distribution as a special case. As seen in the last two chapters, if the true lifetime distribution is chosen to fit the one-shot device data, the MLEs possess less bias and smaller mean square errors. Moreover, the confidence intervals for parameters of interest also maintain nominal levels in most cases. However, these may not hold if a wrong distribution is fitted to the observed data. In this regard, much attention has been paid to model mis-specification analysis and model selection criteria. This chapter examines the effects of model mis-specification on point and interval estimation methods and then describes several statistics useful in model selection. The discussions provided here are primarily from the works carried out by Balakrishnan and Chimitova (2017) and Ling and Balakrishnan (2017).

4.2 Model Mis-Specification Analysis

Ling and Balakrishnan (2017) examined the effects of model mis-specification between gamma and Weibull lifetime distributions on the likelihood inference on mean lifetime and reliability under normal operating conditions based on one-shot device testing data collected in constant-stress accelerated life-tests (CSALTs). Now, let us suppose the lifetimes of one-shot devices arise from distribution A, and the data $\mathbf{z}_A$ are observed. Further, suppose the observed data are wrongly assumed to have come from distribution B. Let $\boldsymbol{\theta}^Q_{A,B}$ be the quasi-MLE (QMLE) (White, 1982) of the model parameter of the mis-specified distribution B

Accelerated Life Testing of One-shot Devices: Data Collection and Analysis, First Edition.
Narayanaswamy Balakrishnan, Man Ho Ling, and Hon Yiu So.

Companion Website: www.wiley.com/go/Balakrishnan/Accelerated_Life_Testing

under the data $\mathbf{z}_A$, given by

$$\theta^Q_{B,A} = \arg\min_{\theta_B} (-\ln(L(\theta_B; \mathbf{z}_A))).$$

To evaluate the effect of model mis-specification, let us denote the bias of the estimator of the parameter of interest ϕ under normal operating conditions $\mathbf{x}_0$ under the wrongly assumed distribution B by $\text{Bias}_B(\phi_A)$, which corresponds to the parameter, $\phi_A(\theta_A)$, under the true distribution. It is given by

$$\text{Bias}_B(\phi_A) = E\left[\hat{\phi}_B\left(\theta^Q_{B,A}\right)\right] - \phi_A(\theta_A). \tag{4.1}$$

Then, the relative bias is given by

$$\text{RB}_B(\phi_A) = \frac{\text{Bias}_B(\phi_A)}{\phi_A(\theta_A)} = \frac{E\left[\hat{\phi}_B\left(\theta^Q_{B,A}\right)\right]}{\phi_A(\theta_A)} - 1. \tag{4.2}$$

Likewise, the model mis-specification effect on the interval estimation can also be investigated, when distribution B is wrongly fitted to data that actually arose from distribution A. To measure the effect of model mis-specification in this case, let $\text{CP}_B(\phi_A)$ denote the coverage probability of the approximate confidence interval, (L_B, U_B), for a particular parameter of interest ϕ under the wrongly assumed distribution B, that is equivalent to $\phi_A(\theta_A)$ under the true distribution A. It can then be expressed as

$$\text{CP}_B(\phi_A) = \Pr(\phi_A(\theta_A) \in (L_B, U_B)), \tag{4.3}$$

where L_B and U_B, respectively, are the lower and upper limits of the confidence interval under the mis-specified distribution B.

4.3 Model Selection

For accurate inference on model parameters and lifetime characteristics of interest, it is important to detect if there is model mis-specification while analyzing one-shot device testing data. In this section, we describe some methods of model selection, based on information criteria and a distance-based test statistic.

4.3.1 Akaike Information Criterion

Akaike information criterion (AIC) (Akaike, 1977, 1981) is a well-known method for measuring the relative quality of fits achieved by probability models, for a given set of data. Suppose distribution A has P model parameters to be estimated. Further, let AIC_A, for example denote the AIC value under distribution A. Then,

$$\text{AIC}_A = 2P - 2\ln(L(\hat{\theta}_A)), \tag{4.4}$$

where $L(\hat{\theta}_A)$ is the likelihood function evaluated with the MLEs of parameters under distribution A. It is evident that AIC makes a trade-off between the goodness-of-fit of the distribution and the simplicity of the model. Therefore, in practice, we select the distribution that is the one with the minimum AIC value from a set of candidate distributions for data.

Now, let AIC_G and AIC_W, for example denote the AIC values under gamma and Weibull distributions, respectively. In our case, for each lifetime distribution, there are $J + 1$ model parameters to be estimated for each of scale and shape parameters, and so AIC_G and AIC_W statistics are given by

$$\text{AIC}_G = 2(2J + 2) - 2\ln(L(\hat{\theta}_G)) \quad \text{and} \quad \text{AIC}_W = 2(2J + 2) - 2\ln(L(\hat{\theta}_W)). \tag{4.5}$$

As the numbers of estimated model parameters are the same for this case when gamma and Weibull lifetime distributions are considered, the AIC criterion will reduce to selecting the lifetime distribution based on the maximized log-likelihood value. When the assumed model does not fit the data well, the log-likelihood value would be expected to become small, and so the preferred model is the one with the smallest AIC value, or equivalently the highest log-likelihood value. We can, therefore, consider a specification test statistic, for the purpose of model selection, as

$$D_{\text{AIC}} = \text{AIC}_G - \text{AIC}_W = 2(\ln(L(\hat{\theta}_W; \mathbf{z})) - \ln(L(\hat{\theta}_G; \mathbf{z}))). \tag{4.6}$$

Naturally, from (4.6), gamma lifetime distribution will be preferred when $D_{\text{AIC}} < 0$, while Weibull lifetime distribution will be preferred when $D_{\text{AIC}} > 0$.

4.3.2 Bayesian Information Criterion

Bayesian information criterion (BIC) (Schwarz, 1978) is another popular measure, with a different penalty for the number of estimated parameters, for model selection. To measure the risk of overfitting the data, both AIC and BIC introduce a penalty function that increases with the number of estimated parameters. Consider distribution A with P estimated parameters for data with n observations. The penalty in AIC is $2P$ (see (4.5)) and is independent of the number of observations, whereas the penalty in BIC is $\ln(n)P$ which increases with the number of estimated parameters P and the number of observations n. So, in a general setting, the BIC value under distribution A is given by

$$\text{BIC}_A = \ln(n)P - 2\ln(L(\hat{\theta}_A)). \tag{4.7}$$

In the case we have focused on, as the number of model parameters in both gamma and Weibull lifetime distributions are the same, the comparison of BIC values is exactly the same as the comparison of the AIC values, which in turn is exactly the same as the comparison of maximized log-likelihood values. Thus,

$$\begin{aligned} D_{\mathrm{BIC}} &= \mathrm{BIC}_G - \mathrm{BIC}_W = 2(\ln(L(\hat{\boldsymbol{\theta}}_W; \mathbf{z})) - \ln(L(\hat{\boldsymbol{\theta}}_G; \mathbf{z}))) \\ &= \mathrm{AIC}_G - \mathrm{AIC}_W = D_{\mathrm{AIC}}. \end{aligned}$$

4.3.3 Distance-Based Test Statistic

In addition, a distance-based test statistic may be used to quantify the distance between the observed and expected numbers of failures in each test group, which may then be used for model validation purpose. Specifically, let M be the test statistic, for a specific assumed distribution, given by

$$M = \max_i |n_i - K_i\hat{F}(\tau_i)|. \tag{4.8}$$

The basis for the form of M in (4.8) is that, when the assumed distribution does not provide a good fit for the observed data, we would expect to observe a large value for it. Note that, in each test group, the number of failures, n_i, has a binomial distribution with sample size K_i and failure probability $\hat{F}(\tau_i)$, and so we can readily calculate the exact p-value as

$$\begin{aligned} p\text{-value} &= \Pr\left(\max_i |n_i - \hat{n}_i| > M\right) \\ &= 1 - \Pr\left(\max_i |n_i - \hat{n}_i| \leq M\right) \\ &= 1 - \Pr\left(|n_i - \hat{n}_i| \leq M \text{ for all } i\right) \\ &= 1 - \prod_{i=1}^{I} \Pr\left(|n_i - \hat{n}_i| \leq M\right) \\ &= 1 - \prod_{i=1}^{I} \Pr\left(-M \leq n_i - \hat{n}_i \leq M\right) \\ &= 1 - \prod_{i=1}^{I} \Pr\left(\hat{n}_i - M \leq n_i \leq \hat{n}_i + M\right) \\ &= 1 - \prod_{i=1}^{I} \left\{ \Pr\left(n_i \leq \hat{n}_i + M\right) - \Pr\left(n_i \leq \hat{n}_i - M - 1\right)\right\} \\ &= 1 - \prod_{i=1}^{I} \left(\sum_{n=\max(0,\lceil \hat{n}_i - M \rceil)}^{\min(K_i, \lceil \hat{n}_i + M - 1 \rceil)} \binom{K_i}{n} \hat{F}(\tau_i)^n (1 - \hat{F}(\tau_i))^{K_i - n} \right), \end{aligned} \tag{4.9}$$

where $\hat{n}_i = K_i\hat{F}(\tau_i)$. If the p-value obtained from (4.9) is small (say, <0.05), then we may conclude that the data do not provide enough evidence to the assumed model.

As with the information-based criteria, we can also propose a specification test statistic as

$$D_M = M_G - M_W, \tag{4.10}$$

for example for selecting between gamma and Weibull distributions. We may conclude that the gamma distribution is the preferred one, based on (4.10), when $D_M < 0$, while the Weibull distribution becomes the preferred one if $D_M > 0$. Furthermore, we may construct confidence intervals $(L, U) = (L_G, U_G)$ when $D_M < 0$ for parameter ϕ_G, and similarly $(L, U) = (L_W, U_W)$ when $D_M > 0$ for parameter ϕ_W. The coverage probabilities of these confidence intervals for ϕ_G and ϕ_W can then be expressed as

$$\begin{aligned} \mathrm{CP}(\phi_G) &= \Pr(\phi_G \in (L_G, U_G), D_M < 0) + \Pr(\phi_G \in (L_W, U_W), D_M > 0) \\ &= \Pr(\phi_G \in (L_G, U_G))P_G(D_M < 0) + \Pr(\phi_G \in (L_W, U_W))P_G(D_M > 0) \end{aligned} \tag{4.11}$$

and

$$\begin{aligned} \mathrm{CP}(\phi_W) &= \Pr(\phi_W \in (L_G, U_G), D_M < 0) + \Pr(\phi_W \in (L_W, U_W), D_M > 0) \\ &= \Pr(\phi_W \in (L_G, U_G))P_W(D_M < 0) + \Pr(\phi_W \in (L_W, U_W))P_W(D_M > 0), \end{aligned} \tag{4.12}$$

where $P_G(D_M < 0)$ represents the probability that the gamma distribution is correctly selected, while $P_W(D_M > 0)$ correspondingly represents the probability that the Weibull distribution is correctly selected by the use of the specification test D_M in (4.10). Here, $P_G(D_M < 0)$ and $P_W(D_M > 0)$ may be regarded as the power of test D_M in correctly identifying gamma and Weibull distributions, respectively. Note that if the gamma distribution is correctly identified, we have

$$\mathrm{CP}(\phi_G) = \Pr(\phi_G \in (L_G, U_G)),$$

which is the coverage probability of the confidence interval for ϕ_G. However, if the specification test chooses the Weibull distribution erroneously, instead of the true gamma distribution, we have

$$\mathrm{CP}(\phi_G) = \Pr(\phi_G \in (L_W, U_W)) = \mathrm{CP}_W(\phi_G).$$

The above means that even if the Weibull distribution is wrongly assumed, there is still a chance that the confidence interval constructed under this wrong distribution would cover the true parameter. Moreover, we see that the power of detection of the two distributions by the specification test statistic D_M becomes important in minimizing the negative impact of model mis-specification on the interval estimation. Finally, it is worth mentioning that the specification test statistic D_M may be replaced by any other specification test statistic, say D_{AIC}, in the above described analysis.

4.3.4 Parametric Bootstrap Procedure for Testing Goodness-of-Fit

Balakrishnan and Chimitova [2017] detailed three goodness-of-fit test statistics, including the Cramer–von Mises–Smirnov-type statistic, Chi-square-type statistic, and White's statistic, for one-shot device testing data collected from CSALTs. However, the p-values for these goodness-of-fit test statistics cannot be expressed in closed-forms. Therefore, they presented a parametric bootstrap procedure for testing the goodness-of-fit, which proceeds as follows:

Step 1: Compute the MLE of model parameter $\hat{\theta}$ for an assumed model based on observed data $\mathbf{z}$;
Step 2: Calculate a test statistic, $S(\hat{\theta})$;
Step 3: Generate one-shot device testing data $\mathbf{z}^*$ for the assumed model with $\hat{\theta}$;
Step 4: Compute the MLE of model parameter $\hat{\theta}^*$ for the assumed model based on simulated data $\mathbf{z}^*$;
Step 5: Calculate the test statistic, $S(\hat{\theta}^*)$;
Step 6: Repeat Steps 3–6 N times;
Step 7: Count the number of simulated data with the test statistic larger than that obtained from the observed data, $N_S = \#\{S(\hat{\theta}^*) > S(\hat{\theta})\}$.

A random sample from the distribution of the test statistic under the null hypothesis can be generated by the above parametric bootstrap procedure. Then, the null hypothesis of goodness-of-fit of the assumed model should be rejected when $N_S/N < 0.05$, for example that can be treated as a bootstrap p-value.

4.4 Simulation Studies

For the one-shot device testing data, there is no analytic expression for the QMLE of the parameters of interest. For this reason, to examine model mis-specification effects, we consider CSALTs for one-shot device testing listed in Table 4.1, simulate 1000 Monte Carlo simulations, and then obtain point and interval estimates of mean lifetime and reliability under normal operating conditions under the assumption of both lifetime distributions.

Table 4.2 presents the relative bias and RMSEs of the estimates of mean lifetime and reliability for different sample sizes K and different normal

Table 4.1 CSALTs with one stress factor on one-shot devices for the simulation study.

Test group i	1	2	3	4	5	6	7	8	9
Inspection time τ_i	12	24	36	12	24	36	12	24	36
Stress level x_i	30	30	30	40	40	40	50	50	50

Table 4.2 Relative bias and RMSEs of the estimates of mean lifetime and reliability at some mission times for various choices of $\mathbf{x}_0$ and sample sizes, for gamma lifetime distribution.

		μ_G	$R_G(10)$	$R_G(20)$	$R_G(30)$	$R_G(40)$	$R_G(50)$
Weibull[a)]	K	**57.3975**	**0.8503**	**0.7165**	**0.6020**	**0.5050**	**0.4231**
$RB_W(\phi_G)$	30	0.1613	0.0226	0.0432	0.0482	0.0394	0.0211
	50	0.0548	0.0256	0.0420	0.0446	0.0360	0.0197
	100	0.0183	0.0258	0.0398	0.0428	0.0373	0.0256
RMSE	30	48.3209	0.0657	0.0740	0.0662	0.0600	0.0666
	50	18.0223	0.0530	0.0599	0.0536	0.0502	0.0566
	100	10.3422	0.0413	0.0465	0.0409	0.0367	0.0395
		μ_G	$R_G(10)$	$R_G(20)$	$R_G(30)$	$R_G(40)$	$R_G(50)$
Weibull[b)]	K	**94.6324**	**0.7563**	**0.6564**	**0.5831**	**0.5242**	**0.4748**
$RB_W(\phi_G)$	30	60.8029	0.1089	0.1455	0.1623	0.1665	0.1617
	50	0.6339	0.1162	0.1467	0.1589	0.1605	0.1554
	100	0.2953	0.1183	0.1442	0.1544	0.1566	0.1540
RMSE	30	1.5274e05	0.1288	0.1395	0.1335	0.1206	0.1061
	50	214.6565	0.1171	0.1254	0.1184	0.1063	0.0940
	100	69.0760	0.1061	0.1110	0.1039	0.0933	0.0827
		μ_G	$R_G(10)$	$R_G(20)$	$R_G(30)$	$R_G(40)$	$R_G(50)$
Gamma[a)]	K	**57.3975**	**0.8503**	**0.7165**	**0.6020**	**0.5050**	**0.4231**
$RB_G(\phi_G)$	30	0.0570	−0.0092	−0.0014	−0.0017	−0.0071	−0.0140
	50	0.0176	0.0001	0.0036	0.0002	−0.0071	−0.0158
	100	0.0049	−0.0001	0.0012	−0.0011	−0.0055	−0.0107
RMSE	30	15.0985	0.0796	0.0789	0.0659	0.0570	0.0582
	50	10.6251	0.0600	0.0595	0.0503	0.0461	0.0491
	100	7.0131	0.0435	0.0421	0.0347	0.0311	0.0330
		μ_G	$R_G(10)$	$R_G(20)$	$R_G(30)$	$R_G(40)$	$R_G(50)$
Gamma[b)]	K	**94.6324**	**0.7563**	**0.6564**	**0.5831**	**0.5242**	**0.4748**
$RB_G(\phi_G)$	30	0.1051	−0.0241	−0.0106	−0.0063	−0.0082	−0.0141
	50	0.0372	−0.0012	0.0068	0.0073	0.0035	−0.0027
	100	0.0130	−0.0008	0.0028	0.0025	0.0002	−0.0033
RMSE	30	40.2380	0.1559	0.1484	0.1339	0.1171	0.1010
	50	26.7347	0.1197	0.1141	0.1018	0.0883	0.0763
	100	17.0971	0.0878	0.0821	0.0721	0.0619	0.0530

a) $\mathbf{x}_0 = 25$.
b) $\mathbf{x}_0 = 15$.

operating conditions $\mathbf{x}_0$ when gamma distribution with $\boldsymbol{\theta}_G = (a_0, a_1, b_0, b_1) = (-1.7, 0.07, 7, -0.12)$ is mis-specified by Weibull distribution. Table 4.3 similarly presents the results when Weibull distribution with $\boldsymbol{\theta}_W = (r_0, r_1, s_0, s_1) = (-0.6, 0.03, 5.3, -0.05)$ is mis-specified by gamma distribution. We observe from these results that, when $\mathbf{x}_0 = 25$, the relative bias of the estimates of mean lifetime and reliability are small for all sample sizes, revealing that the consequence of model mis-specification on point estimation is not serious when normal operating conditions are close to the elevated stress levels used in the CSALT experiment. We additionally observe that the accuracy of estimation is low, when the stress levels used are far from normal operating conditions, which suggests that having test stress levels not too far from normal operating conditions would help mitigate the effect of model mis-specification. Furthermore, if the value of normal operating conditions $\mathbf{x}_0$ is fixed, the relative bias of the estimates of reliability for any fixed time t is approximately stable for all three sample sizes considered. In addition, we observe that $\max_i |F_G(\tau_i; \boldsymbol{\theta}_G) - F_W(\tau_i; \boldsymbol{\theta}^Q_{W,G})| = 0.0227$ and $\max_i |F_W(\tau_i; \boldsymbol{\theta}_W) - F_G(\tau_i; \boldsymbol{\theta}^Q_{G,W})| = 0.0216$, which reveal that the models actually fit the cumulative distribution values for each test group very well. Also, when the fitted model is correct, the results show that the relative bias of the estimates of reliability is small. Surprisingly, the relative bias of the estimates of mean lifetime is large when Weibull distribution is the fitted model, no matter whether the true distribution is Weibull or gamma.

A simulation study is also carried out for evaluating the power of detection of the two distributions for the AIC-based and the distance-based test statistics, for different choices of K. These results are presented in Table 4.4, which show that the distance-based test statistic is slightly better than the AIC-based statistic for detecting mis-specification when gamma distribution is the true distribution, while the AIC-based statistic performs much better when Weibull distribution is the true distribution. In practice, the true distribution will be unknown, and so we recommend the AIC-based statistic for the specification test D_{AIC} in (4.6).

Next, we evaluate the performance of approximate 95% confidence intervals for mean lifetime and reliability with and without the use of the specification tests, D_{AIC} in (4.6) and D_M in (4.10). Tables 4.5–4.8 present the coverage probabilities and the average widths of 95% confidence intervals for the parameters of interest with and without the use of a specification test for model selection. We observe that the coverage probabilities under the true distribution maintain the nominal level. Also, these results show that the model mis-specification is not serious only when gamma distribution is mis-specified as Weibull distribution in the case of small sample sizes. In general, the confidence intervals are seen to possess deflated

Table 4.3 Relative bias and RMSEs of the estimates of mean lifetime and reliability at some mission times for various choices of $\mathbf{x}_0$ and sample sizes, for Weibull lifetime distribution

		μ_W	$R_W(10)$	$R_W(20)$	$R_W(30)$	$R_W(40)$	$R_W(50)$
Gamma[a)]	K	54.4649	0.8769	0.7454	0.6246	0.5182	0.4266
$RB_G(\phi_W)$	30	0.0448	−0.0307	−0.0371	−0.0443	−0.0506	−0.0533
	50	0.0411	−0.0305	−0.0383	−0.0430	−0.0443	−0.0412
	100	0.0208	−0.0243	−0.0344	−0.0400	−0.0413	−0.0381
RMSE	30	12.2623	0.0798	0.0815	0.0716	0.0653	0.0658
	50	9.3396	0.0674	0.0678	0.0578	0.0506	0.0495
	100	6.2349	0.0465	0.0490	0.0431	0.0381	0.0367
		μ_W	$R_W(10)$	$R_W(20)$	$R_W(30)$	$R_W(40)$	$R_W(50)$
Gamma[b)]	K	**102.1655**	**0.8654**	**0.7692**	**0.6893**	**0.6209**	**0.5613**
$RB_G(\phi_W)$	30	−0.0213	−0.1178	−0.1238	−0.1276	−0.1323	−0.1379
	50	−0.0325	−0.1149	−0.1247	−0.1293	−0.1324	−0.1347
	100	−0.0654	−0.0992	−0.1130	−0.1195	−0.1231	−0.1251
RMSE	30	32.0550	0.1799	0.1742	0.1603	0.1447	0.1302
	50	23.7806	0.1577	0.1527	0.1397	0.1252	0.1117
	100	16.9210	0.1197	0.1192	0.1104	0.0998	0.0895
		μ_W	$R_W(10)$	$R_W(20)$	$R_W(30)$	$R_W(40)$	$R_W(50)$
Weibull[a)]	K	**54.4649**	**0.8769**	**0.7454**	**0.6246**	**0.5182**	**0.4266**
$RB_W(\phi_W)$	30	0.1199	−0.0070	−0.0001	0.0006	−0.0055	−0.0153
	50	0.0820	−0.0064	−0.0012	0.0011	0.0001	−0.0026
	100	0.0232	−0.0008	0.0017	0.0017	−0.0008	−0.0048
RMSE	30	27.4570	0.0616	0.0673	0.0608	0.0598	0.0698
	50	16.6614	0.0507	0.0538	0.0467	0.0446	0.0522
	100	8.1583	0.0333	0.0370	0.0329	0.0318	0.0372
		μ_W	$R_W(10)$	$R_W(20)$	$R_W(30)$	$R_W(40)$	$R_W(50)$
Weibull[b)]	K	**102.1655**	**0.8654**	**0.7692**	**0.6893**	**0.6209**	**0.5613**
$RB_W(\phi_W)$	30	7.3648	−0.0182	−0.0064	0.0008	0.0028	0.0000
	50	0.6409	−0.0151	−0.0068	−0.0010	0.0017	0.0019
	100	0.0985	−0.0034	0.0011	0.0035	0.0038	0.0026
RMSE	30	1.5572e04	0.0962	0.1016	0.0968	0.0880	0.0798
	50	563.5662	0.0801	0.0839	0.0782	0.0693	0.0611
	100	40.0980	0.0530	0.0579	0.0545	0.0484	0.0425

a) $\mathbf{x}_0 = 25$.
b) $\mathbf{x}_0 = 15$.

Table 4.4 Power of detection of the two distributions by the AIC and distance-based test statistic M.

True model	$K = 30$		$K = 50$		$K = 100$		$K = 200$	
	AIC	M	AIC	M	AIC	M	AIC	M
Gamma	0.519	0.563	0.523	0.555	0.594	0.606	0.733	0.680
Weibull	0.679	0.530	0.717	0.558	0.752	0.580	0.791	0.637

Table 4.5 Coverage probability (CP) and average width (AW) of 95% confidence intervals for mean lifetime and reliability at four mission times at $\mathbf{x}_0 = 25$, for true (gamma) and mis-specified (Weibull) lifetime distributions.

	$K = 30$		$K = 50$		$K = 100$	
CP	**Gamma**	**Weibull**	**Gamma**	**Weibull**	**Gamma**	**Weibull**
μ_G	0.956	0.937	0.933	0.919	0.940	0.933
$R_G(10)$	0.948	0.961	0.961	0.953	0.941	0.922
$R_G(20)$	0.952	0.952	0.950	0.946	0.946	0.903
$R_G(30)$	0.948	0.948	0.942	0.942	0.949	0.904
$R_G(40)$	0.959	0.960	0.951	0.930	0.954	0.922
AW	**Gamma**	**Weibull**	**Gamma**	**Weibull**	**Gamma**	**Weibull**
μ_G	56.757	278.845	39.677	65.058	26.809	37.673
$R_G(10)$	0.313	0.249	0.237	0.191	0.167	0.135
$R_G(20)$	0.289	0.252	0.224	0.198	0.160	0.141
$R_G(30)$	0.245	0.224	0.189	0.175	0.134	0.124
$R_G(40)$	0.226	0.224	0.174	0.174	0.122	0.123

coverage probabilities and the consequence of model mis-specification on interval estimation is seen to be serious when the stress levels are far from normal operating conditions, especially in the case of large sample sizes. However, even when the coverage probability under a mis-specified distribution is low in the case of large sample sizes, with the use of a specification test, the wrong distribution is unlikely to be selected between the two distributions, as seen from the corresponding power values in Table 4.4. Consequently, the coverage probabilities can be improved considerably and be brought close to the nominal level, with the use of a model specification test in (4.6) or (4.10).

Table 4.6 Coverage probability (CP) and average width (AW) of 95% confidence intervals for mean lifetime and reliability at four mission times at $\mathbf{x}_0 = 25$, for true (Weibull) and mis-specified (gamma) lifetime distributions.

	$K = 30$		$K = 50$		$K = 100$	
CP	**Weibull**	**Gamma**	**Weibull**	**Gamma**	**Weibull**	**Gamma**
μ_W	0.934	0.944	0.941	0.949	0.950	0.959
$R_W(10)$	0.951	0.931	0.936	0.909	0.948	0.907
$R_W(20)$	0.952	0.931	0.948	0.906	0.946	0.893
$R_W(30)$	0.952	0.927	0.952	0.898	0.953	0.876
$R_W(40)$	0.962	0.930	0.949	0.916	0.955	0.880
AW	**Weibull**	**Gamma**	**Weibull**	**Gamma**	**Weibull**	**Gamma**
μ_W	109.444	47.745	56.859	35.904	31.380	24.158
$R_W(10)$	0.254	0.318	0.194	0.242	0.135	0.168
$R_W(20)$	0.261	0.297	0.203	0.232	0.145	0.164
$R_W(30)$	0.232	0.253	0.181	0.196	0.129	0.138
$R_W(40)$	0.227	0.231	0.175	0.178	0.125	0.125

Table 4.7 Average width of 95% confidence intervals for mean lifetime and reliability at four mission times at $\mathbf{x}_0 = 25$ with and without the use of specification tests.

	$K = 30$			$K = 50$			$K = 100$		
ϕ_G	None	AIC	M	None	AIC	M	None	AIC	M
μ_G	278	165	103	65	52	49	37	31	30
$R_G(10)$	0.249	0.285	0.287	0.191	0.217	0.218	0.135	0.155	0.155
$R_G(20)$	0.252	0.272	0.274	0.198	0.212	0.213	0.141	0.153	0.153
$R_G(30)$	0.224	0.235	0.236	0.175	0.182	0.183	0.124	0.130	0.130
$R_G(40)$	0.224	0.223	0.224	0.174	0.173	0.174	0.123	0.121	0.122
$\boldsymbol{\phi_W}$	**None**	**AIC**	$\boldsymbol{M}$	**None**	**AIC**	$\boldsymbol{M}$	**None**	**AIC**	$\boldsymbol{M}$
μ_W	47	75	81	35	51	48	24	29	28
$R_W(10)$	0.318	0.278	0.286	0.242	0.209	0.216	0.168	0.144	0.150
$R_W(20)$	0.297	0.274	0.279	0.232	0.212	0.217	0.164	0.151	0.154
$R_W(30)$	0.253	0.239	0.243	0.196	0.185	0.188	0.138	0.131	0.133
$R_W(40)$	0.231	0.227	0.229	0.178	0.175	0.176	0.125	0.125	0.125

Table 4.8 Coverage probability of mean lifetime and reliability at four mission times for two choices of $\mathbf{x}_0$ with and without the use of specification tests.

	$K = 30$			$K = 50$			$K = 100$		
ϕ_G[a)]	None	AIC	M	None	AIC	M	None	AIC	M
μ_G	0.937	0.946	0.946	0.919	0.928	0.924	0.933	0.936	0.935
$R_G(10)$	0.961	0.955	0.954	0.953	0.955	0.954	0.922	0.937	0.936
$R_G(20)$	0.952	0.949	0.951	0.946	0.943	0.943	0.903	0.936	0.933
$R_G(30)$	0.948	0.947	0.946	0.942	0.939	0.942	0.904	0.936	0.932
$R_G(40)$	0.960	0.962	0.958	0.930	0.939	0.943	0.922	0.939	0.941
ϕ_G[b)]	**None**	**AIC**	$\boldsymbol{M}$	**None**	**AIC**	$\boldsymbol{M}$	**None**	**AIC**	$\boldsymbol{M}$
μ_G	0.966	0.957	0.951	0.956	0.946	0.937	0.973	0.958	0.954
$R_G(10)$	0.982	0.958	0.957	0.937	0.947	0.943	0.785	0.897	0.887
$R_G(20)$	0.974	0.960	0.959	0.912	0.938	0.933	0.729	0.877	0.860
$R_G(30)$	0.963	0.965	0.963	0.877	0.918	0.916	0.669	0.863	0.849
$R_G(40)$	0.943	0.957	0.957	0.859	0.911	0.909	0.606	0.836	0.823
	$\boldsymbol{K = 30}$			$\boldsymbol{K = 50}$			$\boldsymbol{K = 100}$		
ϕ_W[a)]	**None**	**AIC**	$\boldsymbol{M}$	**None**	**AIC**	$\boldsymbol{M}$	**None**	**AIC**	$\boldsymbol{M}$
μ_W	0.944	0.929	0.932	0.949	0.942	0.942	0.959	0.950	0.953
$R_W(10)$	0.931	0.943	0.942	0.909	0.919	0.919	0.907	0.933	0.926
$R_W(20)$	0.931	0.943	0.944	0.906	0.928	0.927	0.893	0.927	0.921
$R_W(30)$	0.927	0.938	0.942	0.898	0.928	0.930	0.876	0.932	0.914
$R_W(40)$	0.930	0.949	0.941	0.916	0.929	0.935	0.880	0.937	0.921
ϕ_W[b)]	**None**	**AIC**	$\boldsymbol{M}$	**None**	**AIC**	$\boldsymbol{M}$	**None**	**AIC**	$\boldsymbol{M}$
μ_W	0.910	0.912	0.905	0.919	0.928	0.925	0.892	0.926	0.914
$R_W(10)$	0.858	0.898	0.893	0.798	0.874	0.868	0.766	0.890	0.862
$R_W(20)$	0.884	0.910	0.908	0.829	0.890	0.883	0.774	0.890	0.864
$R_W(30)$	0.895	0.915	0.916	0.833	0.892	0.883	0.776	0.892	0.864
$R_W(40)$	0.891	0.918	0.914	0.832	0.889	0.882	0.768	0.893	0.856

a) $\mathbf{x}_0 = 25$.
b) $\mathbf{x}_0 = 15$.

4.5 Case Study with R Codes

In this section, we consider serial sacrifice data in Table 1.7. Here, we ignore the information about disease categories and combine the two categories into

Table 4.9 Modified serial sacrifice data, from Table 1.7, on the presence or absence of diseases.

Test group	Sacrifice time (d)	γ-radiation	Number of mice sacrificed	Number of mice with disease(s)
1	100	No	72	14
2	200	No	65	25
3	300	No	63	45
4	400	No	40	32
5	500	No	39	38
6	600	No	33	32
7	700	No	49	49
8	100	Yes	67	13
9	200	Yes	68	32
10	300	Yes	66	53
11	400	Yes	43	43
12	500	Yes	39	39
13	600	Yes	31	31
14	700	Yes	29	29

one. In other words, we are concerned with the data on the presence or absence of diseases. The modified data are presented in Table 4.9. The R codes for the estimation for exponential, gamma, and Weibull lifetime distributions are presented in Table 2.12. Finally, the outputs for the three distributions are reported in Table 4.10. A small p-value is observed for exponential distribution, which means the data provide enough evidence to support the exponential distribution

Table 4.10 R outputs for the log-likelihood, the AIC and the distance-based test statistic in (4.10) for exponential (Exp), gamma (Gam), and Weibull (Wei) lifetime distributions.

Distribution (MLE)	log-likelihood[a)]	AIC	M[b)]	p-Value[c)]
Exp(−5.531, 0.227)	−294.178	592.357	13.265	0.003
Gam(0.946, 0.400, 4.556,−0.584)	−266.012	540.024	5.639	0.399
Wei(0.549, 0.247, 5.614,−0.176)	−263.166	534.332	3.580	0.819

a) M$log-likelihood;
b) M$M;
c) M$pv

for not fitting the data. On the other hand, we observe that both gamma and Weibull distributions fit the data very well, according to the p-values in Table 4.10. However, when the log-likelihood and the distance-based test statistics between the gamma and Weibull distributions are compared, the Weibull distribution seems to have less discrepancy between the expected number of failures and the observed number of failures in the modified serial sacrifice data in Table 4.9. Therefore, the Weibull distribution is recommended for fitting these data. It is worth noting that the p-value is used to measure the strength of evidence against an assumed lifetime distribution. The strength of evidence is highly dependent on sample sizes. When the sample size is small, the p-value is usually large, implying that it is unlikely to reject any assumed lifetime distribution. Therefore, the use of p-value is beneficial for model selection as long as the sample size is sufficiently large.

5 Robust Inference

5.1 Brief Overview

Maximum likelihood estimation method is well known to be efficient, but also non-robust for data with contamination. To mitigate the seriousness of robustness problem, weighted minimum density power divergence estimator, which is a natural extension of the maximum likelihood estimator, can be used as a robust estimator, as it has a better behavior than the maximum likelihood estimator in case of departures from the assumed model. Moreover, the associated asymptotic theory also facilitates the construction of confidence intervals and hypothesis tests. In this chapter, we present the framework for the development of robust estimators for exponential, gamma, and Weibull lifetime distributions for one-shot devices. Wald-type tests based on them are also presented for one-shot device testing data collected from constant-stress accelerated life-tests (CSALTs). The discussions provided here are primarily from the works carried out by Balakrishnan et al. (2019a, b, 2020a). Interested readers may also refer to a recent overview article on density power divergence approach for one-shot device testing data by Balakrishnan et al. (2020c) for more elaborate details.

5.2 Weighted Minimum Density Power Divergence Estimators

From an asymptotic point of view, it is well known that the maximum likelihood estimator is a best asymptotically normal estimator under some general regularity conditions, but at the same time, it has a poor behavior, in general, with regard to robustness. As a gain in robustness often comes at a price in loss of efficiency,

Accelerated Life Testing of One-shot Devices: Data Collection and Analysis, First Edition.
Narayanaswamy Balakrishnan, Man Ho Ling, and Hon Yiu So.

Companion Website: www.wiley.com/go/Balakrishnan/Accelerated_Life_Testing

some divergence measures result in estimators having good properties in terms of robustness with low loss of efficiency. The density power divergence measure (Basu et al., 1998) has such desirable properties and has been studied for many different problems until now. Ghosh and Basu (2013) introduced the minimum density power divergence estimator which can be considered as a natural extension of the maximum likelihood estimator.

For CSALTs with I test groups, we consider the empirical and theoretical probability vectors

$$\hat{\boldsymbol{p}}_i = [\hat{F}_i, \hat{R}_i], \quad i = 1, 2, \dots, I, \tag{5.1}$$

and

$$\boldsymbol{\pi}_i(\boldsymbol{\theta}) = [F_i(\boldsymbol{\theta}), R_i(\boldsymbol{\theta})], \quad i = 1, 2, \dots, I, \tag{5.2}$$

where $\hat{F}_i = n_i/K_i, \hat{R}_i = 1 - n_i/K_i, F_i(\boldsymbol{\theta}) = F(\tau_i; \boldsymbol{\Psi}_i), R_i(\boldsymbol{\theta}) = R(\tau_i; \boldsymbol{\Psi}_i)$, and $\boldsymbol{\Psi}_i = \boldsymbol{\Psi}(\mathbf{x}_i; \boldsymbol{\theta})$.

The Kullback–Leibler divergence measure between $\hat{\boldsymbol{p}}_i$ and $\boldsymbol{\pi}_i(\boldsymbol{\theta})$ is given by

$$d_{\mathrm{KL}}(\hat{\boldsymbol{p}}_i, \boldsymbol{\pi}_i(\boldsymbol{\theta})) = \hat{F}_i \ln\left(\frac{\hat{F}_i}{F_i(\boldsymbol{\theta})}\right) + \hat{R}_i \ln\left(\frac{\hat{R}_i}{R_i(\boldsymbol{\theta})}\right), \tag{5.3}$$

and similarly, the weighted Kullback–Leibler divergence measure for all devices is given by

$$\begin{aligned}\sum_{i=1}^{I} \frac{K_i}{K} d_{\mathrm{KL}}(\hat{\boldsymbol{p}}_i, \boldsymbol{\pi}_i(\boldsymbol{\theta})) &= \sum_{i=1}^{I} \frac{K_i}{K} \left[\hat{F}_i \ln\left(\frac{\hat{F}_i}{F_i(\boldsymbol{\theta})}\right) + \hat{R}_i \ln\left(\frac{\hat{R}_i}{R_i(\boldsymbol{\theta})}\right)\right] \\ &= \frac{1}{K}(s - \ln(L(\boldsymbol{\theta}; \mathbf{z}))),\end{aligned} \tag{5.4}$$

where $K = \sum_{i=1}^{I} K_i$ is the total number of devices, and s is a constant not dependent on $\boldsymbol{\theta}$. Hence, the maximum likelihood estimator of $\boldsymbol{\theta}$ can also be defined as the minimization of the weighted Kullback–Leibler divergence measure in (5.4), that is

$$\hat{\boldsymbol{\theta}} = \arg\min_{\boldsymbol{\theta} \in \boldsymbol{\Theta}} \sum_{i=1}^{I} \frac{K_i}{K} d_{\mathrm{KL}}(\hat{\boldsymbol{p}}_i, \boldsymbol{\pi}_i(\boldsymbol{\theta})) = \arg\min_{\boldsymbol{\theta} \in \boldsymbol{\Theta}} \frac{1}{K}(s - \ln(L(\boldsymbol{\theta}; \mathbf{z}))). \tag{5.5}$$

From the above expression, we can think of defining an estimator minimizing any distance or divergence between the probability vectors $\hat{\boldsymbol{p}}_i$ and $\boldsymbol{\pi}_i(\boldsymbol{\theta})$. There are many different divergence measures known in the literature (Basu et al., 2011; Pardo, 2006), and some are quite suitable to define estimators with good properties in terms of efficiency as well as robustness.

In order to extend the maximum likelihood estimator, we define the weighted density power divergence between $\hat{\mathbf{p}}_i$ and $\boldsymbol{\pi}_i(\boldsymbol{\theta})$, with tuning parameter ω, as follows:

$$\sum_{i=1}^{I} \frac{K_i}{K} d_\omega(\hat{\mathbf{p}}_i, \boldsymbol{\pi}_i(\boldsymbol{\theta})), \tag{5.6}$$

where

$$\begin{aligned} d_\omega(\hat{\mathbf{p}}_i, \boldsymbol{\pi}_i(\boldsymbol{\theta})) = {} & (F_i^{\omega+1}(\boldsymbol{\theta}) + R_i^{\omega+1}(\boldsymbol{\theta})) - \frac{\omega+1}{\omega}(\hat{F}_i F_i^{\omega}(\boldsymbol{\theta}) + \hat{R}_i R_i^{\omega}(\boldsymbol{\theta})) \\ & + \frac{1}{\omega}(\hat{F}_i^{\omega+1} + \hat{R}_i^{\omega+1}). \end{aligned} \tag{5.7}$$

It is evident that the last term $\frac{1}{\omega}(\hat{F}_i^{\omega+1} + \hat{R}_i^{\omega+1})$ does not depend on $\boldsymbol{\theta}$ and so does not have any role in the minimization of the weighted density power divergence measure. Thus, we can define

$$d_\omega^*(\hat{\mathbf{p}}_i, \boldsymbol{\pi}_i(\boldsymbol{\theta})) = (F_i^{\omega+1}(\boldsymbol{\theta}) + R_i^{\omega+1}(\boldsymbol{\theta})) - \frac{\omega+1}{\omega}(\hat{F}_i F_i^{\omega}(\boldsymbol{\theta}) + \hat{R}_i R_i^{\omega}(\boldsymbol{\theta})) \tag{5.8}$$

and further define the weighted minimum density power divergence estimator (WMDPDE) as follows:

$$\hat{\boldsymbol{\theta}}_\omega = \arg\min_{\boldsymbol{\theta}\in\Theta} \sum_{i=1}^{I} \frac{K_i}{K} d_\omega^*(\hat{\mathbf{p}}_i, \boldsymbol{\pi}_i(\boldsymbol{\theta})). \tag{5.9}$$

The WMDPDE can then be determined from the following estimating equations

$$\sum_{i=1}^{I} (K_i F_i(\boldsymbol{\theta}) - n_i)(F_i^{\omega-1}(\boldsymbol{\theta}) + R_i^{\omega-1}(\boldsymbol{\theta})) \frac{\partial F_i(\boldsymbol{\theta})}{\partial \boldsymbol{\theta}} = \mathbf{0} \tag{5.10}$$

for $\omega \geq 0$.

The first-order derivatives of $F_i(\boldsymbol{\theta}) = F(\tau_i; \boldsymbol{\theta})$, with respect to model parameters for exponential, gamma, and Weibull lifetime distributions, are all given presented in Section 2.4.1. Moreover, when $\omega = 0$, the density power divergence is the Kullback–Leibler divergence (see, for example, Kullback and Leibler 1951; Basu et al. 1998), that is

$$d_{\omega=0}(\hat{\mathbf{p}}_i, \boldsymbol{\pi}_i(\boldsymbol{\theta})) = \lim_{\omega\to 0^+} d_\omega(\hat{\mathbf{p}}_i, \boldsymbol{\pi}_i(\boldsymbol{\theta})) = d_{\mathrm{KL}}(\hat{\mathbf{p}}_i, \boldsymbol{\pi}_i(\boldsymbol{\theta})).$$

This result implies that the WMDPDE in (5.9) contains the maximum likelihood estimator as a particular case when $\omega = 0$.

5.3 Asymptotic Distributions

Let $\boldsymbol{\theta}^0$ be the true value of the parameter $\boldsymbol{\theta}$. Then, the asymptotic distribution of the WMDPDE $\hat{\boldsymbol{\theta}}_\omega$ in (5.9), under one-shot device testing data, is given by

$$\sqrt{K}(\hat{\boldsymbol{\theta}}_\omega - \boldsymbol{\theta}^0) \xrightarrow[K\to\infty]{\mathcal{L}} N(\mathbf{0}, \mathbf{J}_\omega^{-1}(\boldsymbol{\theta}^0)\mathbf{K}_\omega(\boldsymbol{\theta}^0)\mathbf{J}_\omega^{-1}(\boldsymbol{\theta}^0)),$$

where

$$\mathbf{J}_\omega(\theta) = \frac{1}{K}\sum_{i=1}^{I} K_i \left(F_i^{\omega-1}(\theta) + R_i^{\omega-1}(\theta)\right) \left(\frac{\partial F_i(\theta)}{\theta}\right) \left(\frac{\partial F_i(\theta)}{\theta'}\right), \tag{5.11}$$

$$\mathbf{K}_\omega(\theta) = \frac{1}{K}\sum_{i=1}^{I} K_i F_i(\theta) R_i(\theta) \left(F_i^{\omega-1}(\theta) + R_i^{\omega-1}(\theta)\right)^2 \left(\frac{\partial F_i(\theta)}{\theta}\right) \left(\frac{\partial F_i(\theta)}{\theta'}\right). \tag{5.12}$$

Interested readers may refer to Ghosh and Basu (2013) for pertinent details. As $\hat{\theta}_{\omega=0}$ is the maximum likelihood estimator of θ, it is also evident that

$$\mathbf{J}_{\omega=0}(\theta) = \mathbf{K}_{\omega=0}(\theta) = I_{\text{obs}}(\theta),$$

implying that the asymptotic variance of $\hat{\theta}_{\omega=0}$ becomes the inverse of the Fisher information matrix, $I_{\text{obs}}(\theta)$, in (2.25), which is well known in the classical asymptotic theory of maximum likelihood estimator.

5.4 Robust Wald-type Tests

Based on the asymptotic distribution of the WMDPDE, we can define Wald-type tests for testing a composite null hypothesis about a function of model parameter θ.

Let us consider the function $\boldsymbol{m} : \mathbb{R}^S \to \mathbb{R}^r$, where $r \leq S$. Then, $\boldsymbol{m}(\theta) = \mathbf{0}_r$ represents a composite null hypothesis. We assume that the $S \times r$ matrix

$$\boldsymbol{M}(\theta) = \frac{\partial \boldsymbol{m}'(\theta)}{\partial \theta}$$

exists and is continuous in θ, with rank $\boldsymbol{M}(\theta) = r$. In the context of one-shot device testing under multiple factors, we may be interested in checking whether there is a significant relationship between the jth stress factor and the lifetime of devices under gamma distribution, for example. In this case, we can set $r = 2, \boldsymbol{m}(\theta) = (a_j, b_j)'$ and

$$\boldsymbol{M}(\theta) = \begin{pmatrix} 0 \cdots & \overbrace{1}^{j+1} & \cdots & \overbrace{0}^{j+J+2} & \cdots & \overbrace{0}^{2J+2} \\ 0 \cdots & 0 & \cdots & 1 & \cdots & 0 \end{pmatrix}.$$

For testing

$$H_0 : \theta \in \boldsymbol{\Theta}_0 \quad \text{against} \quad H_1 : \theta \notin \boldsymbol{\Theta}_0, \tag{5.13}$$

where $\boldsymbol{\Theta}_0 = \{\boldsymbol{\theta} \in \boldsymbol{\Theta} : \boldsymbol{m}(\boldsymbol{\theta}) = \mathbf{0}_r\}$, the Wald-type test statistic is given by

$$W_K(\hat{\boldsymbol{\theta}}_\omega) = K\boldsymbol{m}'(\hat{\boldsymbol{\theta}}_\omega)\,(\boldsymbol{M}'(\hat{\boldsymbol{\theta}}_\omega)\boldsymbol{\Sigma}(\hat{\boldsymbol{\theta}}_\omega)\boldsymbol{M}(\hat{\boldsymbol{\theta}}_\omega))^{-1}\boldsymbol{m}(\hat{\boldsymbol{\theta}}_\omega), \tag{5.14}$$

where $\boldsymbol{\Sigma}(\hat{\boldsymbol{\theta}}_\omega) = \mathbf{J}_\omega^{-1}(\hat{\boldsymbol{\theta}})\mathbf{K}_\omega(\hat{\boldsymbol{\theta}})\mathbf{J}_\omega^{-1}(\hat{\boldsymbol{\theta}})$ and $\mathbf{J}_\omega(\boldsymbol{\theta})$ and $\mathbf{K}_\omega(\boldsymbol{\theta})$ are as given in (5.11) and (5.12), respectively.

Then, the asymptotic null distribution of the Wald-type test statistic is a χ^2 distribution with r degrees of freedom, i.e.

$$W_K(\hat{\boldsymbol{\theta}}_\omega) \xrightarrow[K\to\infty]{\mathcal{L}} \chi^2_r.$$

Interested readers may refer to Basu et al. (2016,2018) for general results in relation to Wald-type tests based on WMDPDEs.

Hence, at a significance level of δ, we will reject the null hypothesis when

$$W_K(\hat{\boldsymbol{\theta}}_\omega) > \chi^2_{r,\delta},$$

where $\chi^2_{r,\delta}$ is the upper δ percentage point of χ^2_r distribution.

5.5 Influence Function

An influence function is an important concept in robustness theory, see, for example Hampel et al. (1986) and Balakrishnan et al. (2020c). For any estimator defined in terms of a statistical functional $U(F)$ from the true distribution F, its influence function is defined as

$$\text{IF}(t, U, F) = \lim_{\epsilon\to 0}\frac{U(F_\epsilon) - U(F)}{\epsilon} = \left.\frac{\partial U(F_\epsilon)}{\partial \epsilon}\right|_{\epsilon=0^+}, \tag{5.15}$$

where $F_\epsilon = (1-\epsilon)F + \epsilon\Delta_t$, with ϵ being the contamination proportion and Δ_t being the degenerate distribution at the contamination point t. It is evident that the influence function is a function of t that measures the standardized asymptotic bias (in its first-order approximation) caused by an infinitesimal contamination at point t. The maximum of this influence function over t indicates the extent of bias due to contamination. Thus, an estimator is more robust when its value of the influence function is smaller.

In the context of one-shot device testing data, let us denote the true distribution function of a Bernoulli random variable with an unknown probability success by G_i and having probability mass function g_i, for the ith test group of K_i observations. Similarly, let $F_{i,\theta}$ be the distribution of a Bernoulli random variable having a probability of success $F_i(\boldsymbol{\theta})$, with probability mass function $f_i(x, \boldsymbol{\theta})$, which are related to the model. In vector notation, we consider

$$\boldsymbol{G} = (G_1 \bigotimes \mathbf{1}'_{K_1}, G_2 \bigotimes \mathbf{1}'_{K_2}, \ldots, G_I \bigotimes \mathbf{1}'_{K_I})'$$

and

$$\boldsymbol{F}_{\theta} = (F_{1,\theta} \bigotimes \mathbf{1}'_{K_1}, F_{2,\theta} \bigotimes \mathbf{1}'_{K_2}, \dots, F_{I,\theta} \bigotimes \mathbf{1}'_{K_I})'.$$

We then need to define the statistical functional $\boldsymbol{U}_{\omega}(\boldsymbol{G})$ corresponding to the WMDPDE as the minimizer of the weighted sum of density power divergences between the true and model densities, that is

$$H_{\omega}(\boldsymbol{\theta}) = \sum_{i=1}^{I} \frac{K_i}{K} \left\{ \sum_{x \in \{0,1\}} \left(f_i^{\omega+1}(x, \boldsymbol{\theta}) - \frac{\omega+1}{\omega} f_i^{\omega}(x, \boldsymbol{\theta}) g_i(x) \right) \right\}, \tag{5.16}$$

where $f_i(x, \boldsymbol{\theta}) = xF_i(\boldsymbol{\theta}) + (1-x)R_i(\boldsymbol{\theta})$ and $g_i(x)$ is the probability mass function associated to G_i.

We observe that (5.16) gets minimized at $\boldsymbol{\theta} = \boldsymbol{\theta}^0$ when $g_i(x) \equiv f_i(x, \boldsymbol{\theta})$, implying that the Fisher information matrix consistency of the WMDPDE functional $\boldsymbol{U}_{\omega}(\boldsymbol{G})$ in our model.

To get the influence function of the WMDPDE at $\boldsymbol{F}_{\theta}$ with respect to the kth element of the i_0th test group, we replace $\boldsymbol{\theta}$ in (5.16) with

$$\boldsymbol{\theta}_{\epsilon}^{i_0} = \boldsymbol{U}_{\omega}\,(G_1 \bigotimes \mathbf{1}'_{K_1}, \dots, G_{i_0,\epsilon} \bigotimes \mathbf{1}'_{K_{i_0}}, \dots, G_I \bigotimes \mathbf{1}'_{K_I}),$$

where $G_{i_0,\epsilon}$ is the distribution function associated to the probability mass function

$$g_i(x) = \begin{cases} f_i(x, \boldsymbol{\theta}^0), & \text{if } i \neq i_0, \\ g_{i_0,\epsilon,k}(x), & \text{if } i = i_0, \end{cases}$$

$$g_{i_0,\epsilon,k}(x) = (1-\epsilon) f_i(x, \boldsymbol{\theta}^0) + \epsilon(x\Delta_{t_{i_0,k}} + (1-x)(1-\Delta_{t_{i_0,k}})),$$

and $\Delta_{t_{i_0,k}}$ is the degenerating function at point $t_{i_0,k}$.

Differentiating $g_{i_0,\epsilon,k}(x)$ with respect to ϵ and evaluating this derivative at $\epsilon = 0$, the influence function with respect to the kth element of the i_0th test group is obtained as

$$\begin{aligned} &\mathrm{IF}(t_{i_0,k}, \boldsymbol{U}_{\omega}, \boldsymbol{F}_{\theta^0}) \\ &= \mathbf{J}_{\omega}^{-1}(\boldsymbol{\theta}^0) \left(\frac{K_{i_0}}{K} \right) (F_{i_0}^{\omega-1}(\boldsymbol{\theta}^0) + R_{i_0}^{\omega-1}(\boldsymbol{\theta}^0)) \, (F_{i_0}(\boldsymbol{\theta}^0) - \Delta_{t_{i_0,k}}) \left. \frac{\partial F_{i_0}(\boldsymbol{\theta})}{\boldsymbol{\theta}} \right|_{\boldsymbol{\theta}=\boldsymbol{\theta}^0}. \end{aligned} \tag{5.17}$$

Similarly, the influence function with respect to all the observations is given by

$$\begin{aligned} \mathrm{IF}(\boldsymbol{t}, \boldsymbol{U}_{\omega}, \boldsymbol{F}_{\theta^0}) &= \lim_{\epsilon \to 0} \frac{\boldsymbol{U}(\boldsymbol{G}_{\epsilon,t}) - \boldsymbol{U}(\boldsymbol{G})}{\epsilon} = \left. \frac{\partial \boldsymbol{U}(\boldsymbol{G}_{\epsilon,t})}{\partial \epsilon} \right|_{\epsilon=0^+} \\ &= \mathbf{J}_{\omega}^{-1}(\boldsymbol{\theta}^0) \sum_{i=1}^{I} \left(\frac{K_i}{K} \right) (F_i^{\omega-1}(\boldsymbol{\theta}^0) + R_i^{\omega-1}(\boldsymbol{\theta}^0))(F_i(\boldsymbol{\theta}^0) - \Delta_{t_i}) \left. \frac{\partial F_i(\boldsymbol{\theta})}{\boldsymbol{\theta}} \right|_{\boldsymbol{\theta}=\boldsymbol{\theta}^0}, \end{aligned} \tag{5.18}$$

where $t = (t_{11}, t_{12}, \dots, t_{1K_1}, \dots, t_{I1}, t_{I2}, \dots, t_{IK_I})'$, $G_{\epsilon,t} = (1-\epsilon)G + \epsilon\Delta_t$ and $\Delta_{t_i} = \sum_{k=1}^{K_i} \Delta_{t_{i,k}}$.

5.6 Simulation Studies

In this section, extensive simulation studies are carried out to examine the behavior of the WMDPDEs of model parameters and the robustness of Wald-type tests with $\delta = 0.05$ level of significance for various sample sizes, different values of tuning parameter and different lifetime distributions. Here, we consider CSALTs with $I = 9$ test groups, $\mathbf{x}_i \in \{35, 45, 55\}$ and $\tau \in \{10, 20, 30\}$, for $i = 1, 2, \dots, I$, and tests about the stress factor on the scale parameter. The settings of null and alternative hypotheses for exponential, gamma, and Weibull lifetime distributions are described in Table 5.1.

In addition, we pay special attention to the robustness issue. Hence, under the setting of CSALTs, there are K observations in each cell, and we consider "outlying cells" rather than "outlying observations." A cell which does not follow the underlying distribution will be called an outlying cell or outlier. The strong outliers may lead to the rejection of a model fit even if the rest of the cells fit the model properly. In other cases, even though the cells seem to fit reasonably well the distribution, the outlying cells contribute to an increase in the values of the residuals as well as the divergence measure between the data and the fitted values according to the distribution considered. For this reason, it is very important to have robust estimators as well as robust test statistics in order to avoid the undesirable effects that arise due to the presence of outliers in the data. To evaluate the robustness of the WMDPDEs, we study the behavior of the distribution under the consideration of an outlying cell with smaller $\theta_1 \in \{e_1, b_1, s_1\}$ for the first test group ($i = 1$) in

Table 5.1 Null and alternative hypotheses for exponential, gamma, and Weibull lifetime distributions.

		True parameter under	
Distribution	**Hypotheses**	H_0	H_1
Exponential	$H_0 : e_1 = 0.05$	(e_0, e_1)	$e_1 = 0.03$
	$H_1 : e_1 \neq 0.05$	$= (-6, 0.05)$	
Gamma	$H_0 : b_1 = -0.05$	(a_0, a_1, b_0, b_1)	$b_1 = -0.03$
	$H_1 : b_1 \neq -0.05$	$= (-1.2, 0.04, 4.5, -0.05)$	
Weibull	$H_0 : s_1 = -0.05$	(r_0, r_1, s_0, s_1)	$s_1 = -0.03$
	$H_1 : s_1 \neq -0.05$	$= (-1, 0.03, 6, -0.05)$	

the CSALTs. The reduction of the magnitude of parameter θ_1 of the outlying cell, denoted by $\tilde{\theta}_1$, increases reliability and mean lifetime. If $\tilde{\theta}_1 = \theta_1$, it implies that there is no outlying cell in the data.

Figures 5.1–5.3 present RMSEs of the WMDPDEs of $\theta_1 \in \{e_1, b_1, s_1\}$, empirical significance levels (measured as proportions of test statistics in absolute value exceeding the critical value of 3.84) and empirical powers, based on 1000 Monte Carlo simulations, for various values of the tuning parameter and sample sizes for exponential, gamma, and Weibull lifetime distributions.

For exponential lifetime distribution, in the cases with small sample sizes, it is observed that the maximum likelihood estimator ($\omega = 0$) and the WMDPDEs with small values of tuning parameter present the smallest RMSEs for no ($\tilde{e}_1 = 0.05$) and weak outliers. On the other hand, large values of tuning parameter lead the WMDPDEs to present smallest RMSEs, for medium and strong outliers and large sample sizes. Also, the maximum likelihood estimator of $\boldsymbol{\theta}_E$ is very efficient when there are no outliers, but highly non-robust when there are outliers, especially for large sample sizes. On the other hand, the WMDPDEs with moderate values of the tuning parameter ω exhibit a little loss of efficiency without outliers, but at the same time provide a considerable improvement in terms of robustness in the presence of outliers. Actually, these values of the tuning parameter ω are the most appropriate ones for the estimation of parameters in the one-shot device model based on robustness consideration, bearing in mind that to improve in a considerable way the robustness of the estimators, a small amount of efficiency needs to be compromised.

We observe that the observed levels are quite close to the nominal level of 0.05 in all the cases without outliers ($\tilde{e}_1 = 0.05$). When there is an outlier placed in the first test group, it should be noted that the outlying cell represents one ninth of the total observations. There is a large inflation in the significance level and shrinkage of the power, but for Wald-type tests based on the WMDPDEs with large values of the tuning parameter, the effect of the outlying cell is weaker as compared to those of smaller values of the tuning parameter, including the maximum likelihood estimator ($\omega = 0$). If $\tilde{e}_1$ is apart from e_1, the significance level of Wald-type tests based on the WMDPDEs is not stable around the nominal level, but does get closer as the tuning parameter becomes larger. Also, the power of Wald-type tests based on the WMDPDEs decreases, but more slowly when the tuning parameter becomes larger. The results also show the poor behavior of Wald-type tests based on the maximum likelihood estimator in terms of robustness. The robustness features of Wald-type tests based on the WMDPDEs with large values of the tuning parameter are often better as they maintain both significance level and power in a stable manner when outliers are present in the data. Finally, we observe that the robustness of the WMDPDE seems to increase with increasing ω, but their efficiency in the case

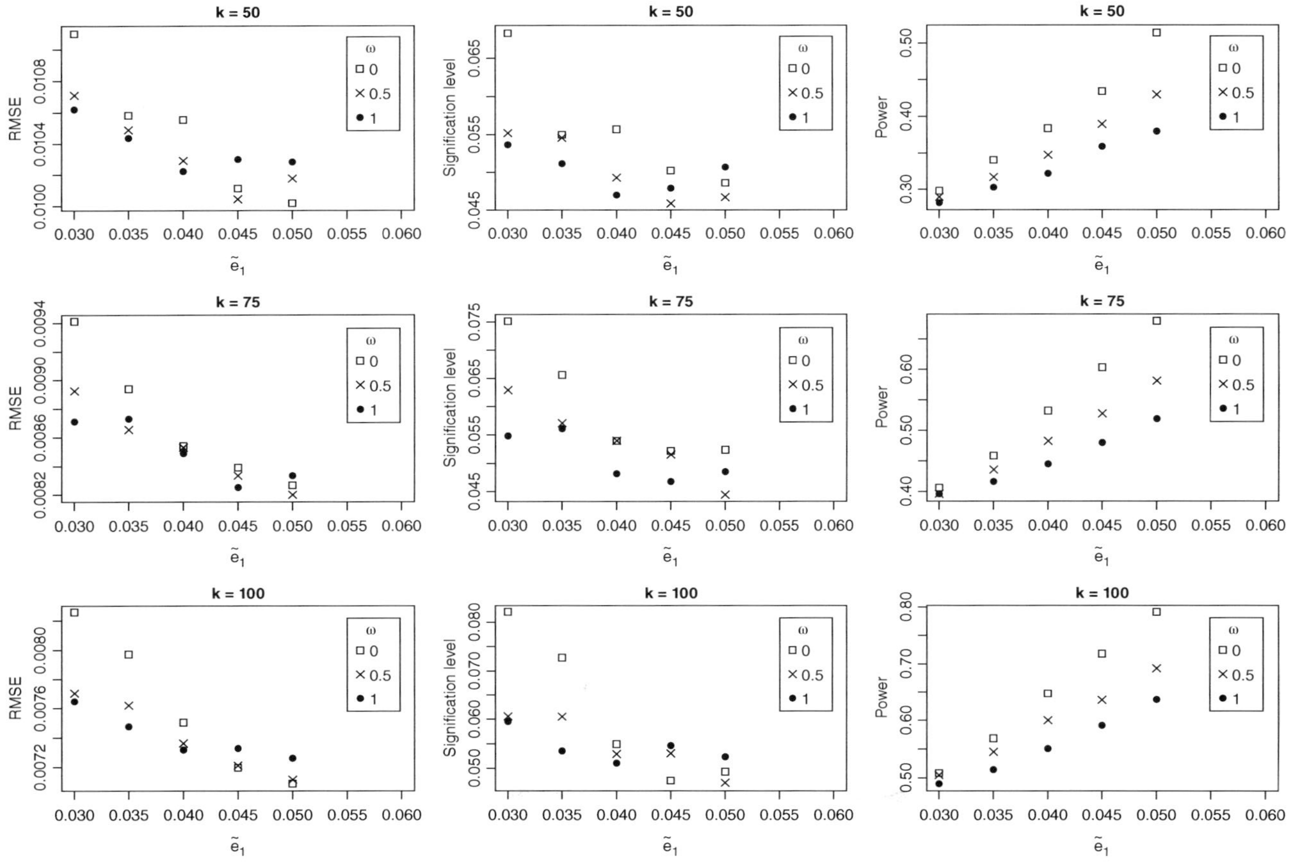

Figure 5.1 RMSE of the WMDPDE of e_1, as well as the significance level and the power of Wald-type tests with $\delta = 0.05$ for various values of the tuning parameter, $\omega \in \{0, 0.5, 1\}$, and sample sizes, $K \in \{50, 75, 100\}$.

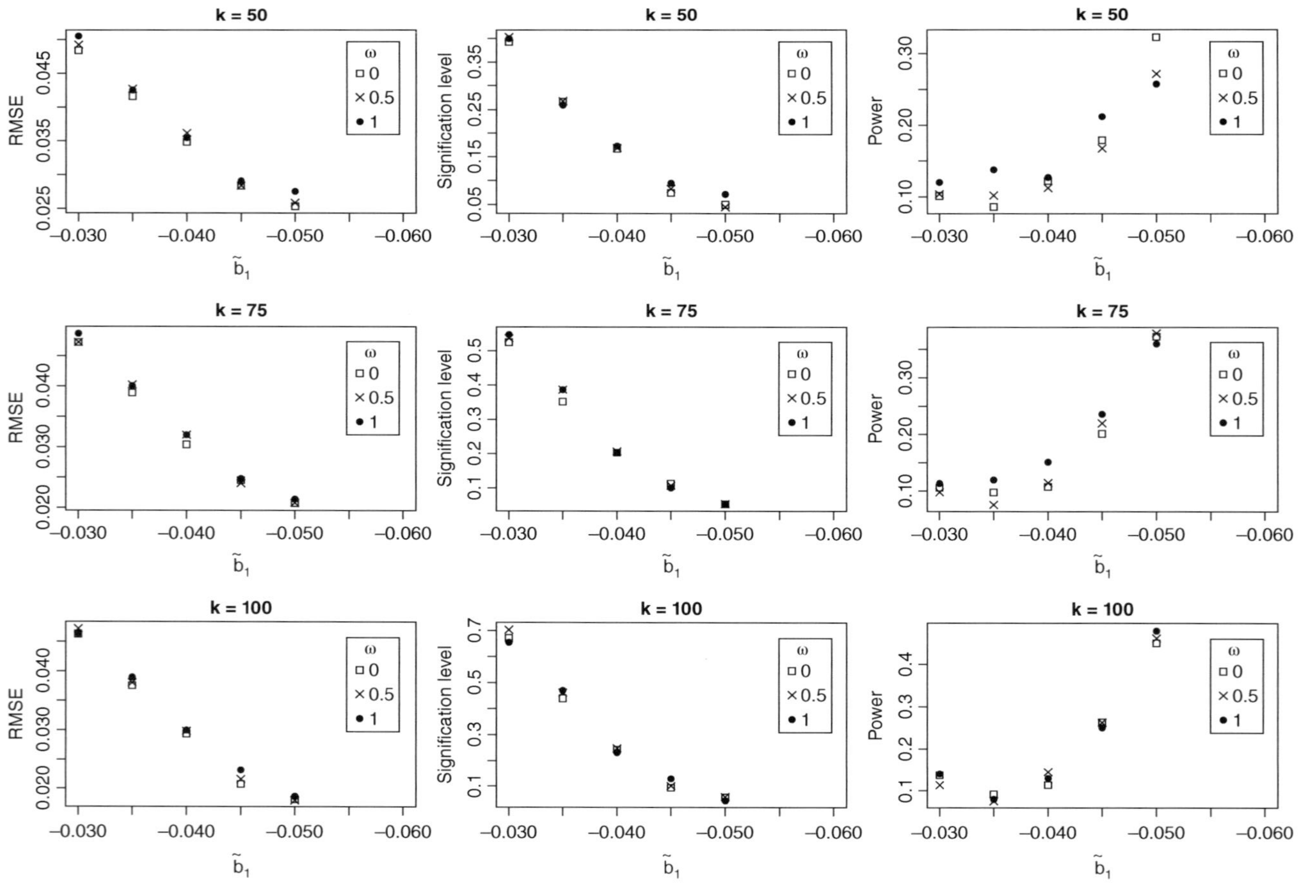

Figure 5.2 RMSE of the WMDPDE of b_1, as well as the significance level and the power of Wald-type tests with $\delta = 0.05$ for various values of the tuning parameter, $\omega \in \{0, 0.5, 1\}$, and sample sizes, $K \in \{50, 75, 100\}$.

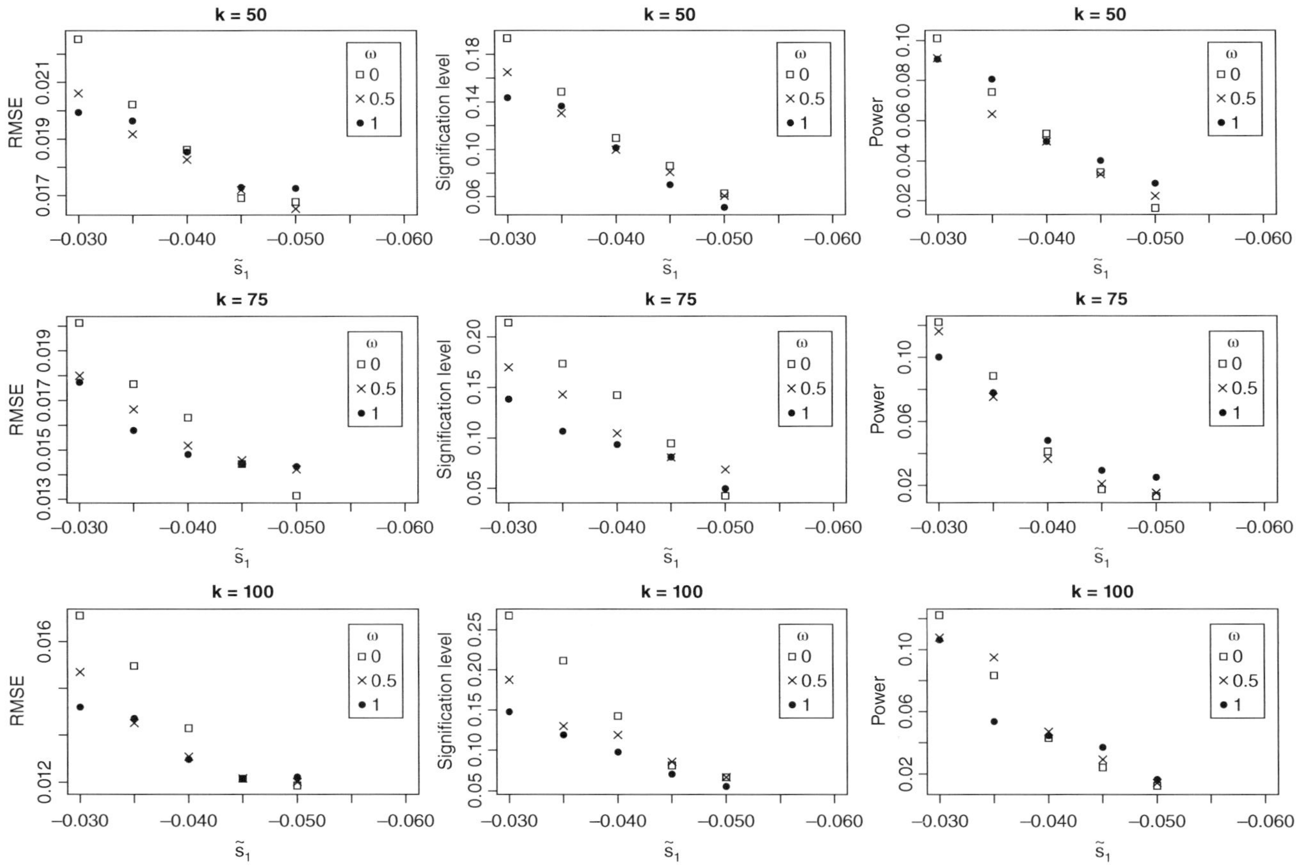

Figure 5.3 RMSE of the WMDPDE of s_1, as well as the significance level and the power of Wald-type tests with $\delta = 0.05$ for various values of the tuning parameter, $\omega \in \{0, 0.5, 1\}$, and sample sizes, $K \in \{50, 75, 100\}$.

of pure data do decrease slightly. We also observe similar results for Weibull lifetime distribution, but the power of Wald-type tests increases when $\tilde{s}_1$ is away from s_1. For the gamma lifetime distribution, the results show similar behavior in RMSE as well as robustness of Wald-type tests based on the maximum likelihood estimator and WMDPDEs. In general, a moderate value of ω is expected to provide the best trade-off for possibly contaminated data, but a data-driven choice of ω would be better and more convenient. Some useful procedures for data-based selection of tuning parameter value are available in the works of Warwick and Jones (2005) and Basak et al. (2020).

Table 5.2 R codes for computing WMDPDEs and Wald-type test statistic values for glass capacitors data presented in Table 1.2 with tuning parameter $\omega = 0.1$..

```
sourc("EM.R")
source("WMDPDE.R")
w<-0.1
#exponential models
p<-n/K
p[p==0]<-0.01
p[p==1]<-0.99
y<-log(-log(1-p))-log(Tau)
A<-matrix(c(rep(1,each=length(p)),X),nrow=length(p))
thetaE0<-solve(t(A)%*%A,t(A)%*%y)
WMD<-WMDPDE(K,n,X,Tau,thetaE0,x0,t0,w,"exp")
m<-matrix(c(0,0,1),nrow=1)
M<-t(m)
Wald<-Robust.Wald(K,X,Tau,WMD$theta.hat,w,m,d=0,M,"exp")
#gamma models
theta0<-c(0,0,0,-thetaE0)
WMD<-WMDPDE(K,n,X,Tau,theta0,x0,t0,w,"gamma")
m<-matrix(c(0,0,0,0,0,1),nrow=1)
M<-t(m)
Wald<-Robust.Wald(K,X,Tau,WMD$theta.hat,w,m,d=0,M,"gamma")
#Weibull models
WMD<-WMDPDE(K,n,X,Tau,theta0,x0,t0,w,"Weibull")
m<-matrix(c(0,0,0,0,0,1),nrow=1)
M<-t(m)
Wald<-Robust.Wald(K,X,Tau,WMD$theta.hat,w,m,d=0,M,"Weibull")
```

5.7 Case Study with R Codes

In this section, we consider the glass capacitors data presented in Table 1.2. The R codes for computing WMDPDEs and Wald-type test statistic for testing $H_0 : e_2 = 0$ against $H_1 : e_2 \neq 0$ (exponential), $H_0 : b_2 = 0$ against $H_1 : b_2 \neq 0$ (gamma) and $H_0 : s_2 = 0$ against $H_1 : s_2 \neq 0$ (Weibull), for various values of the tuning parameter, are displayed in Table 5.2. Finally, the outputs for the WMDPDEs of the model parameters, reliability at various mission times and the mean lifetime under normal operating condition $\mathbf{x}_0 = (1/443, \ln(250))$, along with the corresponding Wald-type test statistic values for exponential, gamma, and Weibull lifetime distributions, are all presented in Tables 5.3–5.5. Note that the predictors, $1/x_1$ and $\ln(x_2)$, are used for temperature (K) and voltage (V) in this study. These results show that the mean lifetime obtained by the maximum likelihood estimator ($\omega = 0$) is generally greater than those obtained from the

Table 5.3 The WMDPDEs of model parameters, reliability at various mission times, $t_1 = 500, t_2 = 800, t_3 = 1000$, and mean lifetime under normal operating condition $\mathbf{x}_0 = (1/443, \ln(250))$ and the Wald-type test statistic values, for various choices of the tuning parameter ω, for exponential lifetime distribution.

ω	$\hat{e}_{0,\omega}$ a)	$\hat{e}_{1,\omega}$ a)	$\hat{e}_{2,\omega}$ a)	$\hat{R}_\omega(t_1)$ b)	$\hat{R}_\omega(t_2)$ b)	$\hat{R}_\omega(t_3)$ b)	$\hat{\mu}_\omega$ c)	W_K d)
0	−23.471	−2669	3.916	0.827	0.738	0.684	2633	5.998
0.1	−22.152	−2766	3.726	0.819	0.726	0.670	2501	5.707
0.2	−21.630	−2830	3.660	0.816	0.722	0.666	2459	5.564
0.3	−21.164	−2872	3.596	0.813	0.718	0.661	2418	5.403
0.4	−20.747	−2898	3.534	0.810	0.714	0.657	2378	5.234
0.5	−20.377	−2913	3.476	0.808	0.711	0.652	2341	5.065
0.6	−20.053	−2921	3.423	0.805	0.707	0.648	2307	4.905
0.7	−19.772	−2923	3.376	0.803	0.704	0.644	2276	4.758
0.8	−19.534	−2922	3.335	0.801	0.701	0.641	2250	4.628
0.9	−19.337	−2920	3.300	0.799	0.698	0.638	2227	4.517
1	−19.179	−2915	3.271	0.797	0.696	0.636	2209	4.426
2	−19.179	−2915	3.271	0.797	0.696	0.636	2209	4.426
3	−20.024	−3204	3.528	0.811	0.715	0.657	2381	5.156
4	−20.410	−3796	3.825	0.824	0.734	0.679	2583	5.882

a) WMD$theta.hat;
b) WMD$R0.hat;
c) WMD$ET0.hat;
d) Wald$W

Table 5.4 The WMDPDEs of model parameters, reliability at various mission times, $t_1 = 500, t_2 = 800, t_3 = 1000$, and mean lifetime under normal operating condition $\mathbf{x}_0 = (1/443, \ln(250))$ and the Wald-type test statistic values, for various choices of the tuning parameter ω, for gamma lifetime distribution.

ω	$\hat{a}_{0,\omega}$ a)	$\hat{a}_{1,\omega}$ a)	$\hat{a}_{2,\omega}$ a)	$\hat{b}_{0,\omega}$ a)	$\hat{b}_{1,\omega}$ a)	$\hat{b}_{2,\omega}$ a)	$\hat{R}_\omega(t_1)$ b)	$\hat{R}_\omega(t_2)$ b)	$\hat{R}_\omega(t_3)$ b)	$\hat{\mu}_\omega$ c)	W_K d)
0	5.021	−1746	0.220	1.130	4730	−1.390	0.702	0.169	0.040	617	0.020
0.1	4.915	−1743	0.235	1.124	4726	−1.385	0.695	0.166	0.039	614	0.019
0.2	4.870	−1745	0.248	1.125	4727	−1.390	0.694	0.161	0.037	612	0.019
0.3	4.884	−1731	0.243	1.149	4673	−1.377	0.691	0.155	0.034	609	0.018
0.4	4.883	−1733	0.247	1.130	4669	−1.376	0.690	0.151	0.032	607	0.018
0.5	4.929	−1755	0.251	1.134	4654	−1.374	0.688	0.146	0.030	604	0.017
0.6	4.835	−1786	0.281	1.266	4678	−1.408	0.689	0.145	0.030	604	0.018
0.7	4.813	−1783	0.287	1.251	4680	−1.411	0.688	0.142	0.028	603	0.018
0.8	4.327	−1935	0.435	1.948	4846	−1.600	0.697	0.151	0.031	608	0.028
0.9	0.775	−2775	1.427	6.222	5679	−2.718	0.707	0.154	0.032	613	0.141
1	0.854	−2804	1.428	6.211	5677	−2.719	0.711	0.152	0.031	613	0.134
2	0.837	−2799	1.429	6.195	5687	−2.720	0.711	0.153	0.031	613	0.135
3	0.885	−2817	1.423	6.188	5693	−2.716	0.710	0.158	0.033	615	0.146
4	0.870	−2818	1.415	6.365	5692	−2.735	0.715	0.174	0.040	622	0.159

a) WMD$theta.hat.
b) WMD$R0.hat.
c) WMD$ET0.hat.
d) Wald$W.

Table 5.5 The WMDPDEs of model parameters, reliability at various mission times, $t_1 = 400, t_2 = 800, t_3 = 1000$, and mean lifetime under normal operating condition $\mathbf{x}_0 = (1/443, \ln(250))$ and the Wald-type test statistic values, fir various choices of the tuning parameter ω, for Weibull lifetime distribution.

ω	$\hat{r}_{0,\omega}$ [a)]	$\hat{r}_{1,\omega}$ [a)]	$\hat{r}_{2,\omega}$ [a)]	$\hat{s}_{0,\omega}$ [a)]	$\hat{s}_{1,\omega}$ [a)]	$\hat{s}_{2,\omega}$ [a)]	$\hat{R}_\omega(t_1)$ [b)]	$\hat{R}_\omega(t_2)$ [b)]	$\hat{R}_\omega(t_3)$ [b)]	$\hat{\mu}_\omega$ [c)]	W_K [d)]
0	3.888	−1653	0.240	6.318	3108	−1.243	0.724	0.079	0.001	589	0.555
0.1	4.067	−1673	0.212	6.197	3091	−1.215	0.715	0.079	0.001	587	0.469
0.2	3.955	−1692	0.240	6.268	3079	−1.222	0.715	0.080	0.001	587	0.493
0.3	3.990	−1814	0.286	6.381	3066	−1.238	0.718	0.075	0.001	586	0.559
0.4	2.245	−1819	0.606	6.834	3065	−1.318	0.727	0.079	0.001	591	0.934
0.5	1.270	−1857	0.800	7.130	3065	−1.371	0.735	0.082	0.001	594	1.278
0.6	1.411	−1860	0.778	7.118	3032	−1.356	0.732	0.075	0.001	591	1.198
0.7	−1.663	−2289	1.519	8.124	3015	−1.529	0.760	0.078	0.001	601	3.367
0.8	−1.093	−2874	1.655	8.060	3103	−1.553	0.765	0.081	0.001	604	3.788
0.9	−1.645	−2925	1.786	8.159	3079	−1.562	0.772	0.064	0.000	601	4.042
1	−1.047	−2927	1.675	8.152	3052	−1.550	0.767	0.069	0.000	601	3.489
2	−4.832	−4528	3.056	9.453	2904	−1.725	0.827	0.031	0.000	608	18.211
3	−6.295	−4630	3.371	9.421	2957	−1.741	0.838	0.026	0.000	609	27.971
4	−6.175	−4595	3.332	9.453	2927	−1.735	0.833	0.027	0.000	608	28.861

a) WMD$theta.hat.
b) WMD$R0.hat.
c) WMD$ET0.hat.
d) Wald$W.

alternative WMDPDEs. In addition, the Wald-type test statistic, with various values of the tuning parameter, for testing the hypothesis that different applied voltages do not affect the scale parameter of the lifetime, shows that the null hypothesis cannot not be rejected at a significance level $\delta = 0.05$ for gamma and Weibull (with small ω) lifetime distributions, but does get rejected for exponential lifetime distribution.

6

Semi-Parametric Models and Inference

6.1 Brief Overview

Exponential, gamma, and Weibull lifetime distributions are fully parametric models. As point and interval estimation on lifetime characteristics may be sensitive to departures from an assumed parametric model, semi-parametric models, that consist of parametric and nonparametric parts, can provide an alternative for fitting one-shot device testing data. In this chapter, we consider proportional hazards models that assume proportional hazard rates and involve a link function that is required as a parametric part to extrapolate the data observed from elevated stress levels to estimate lifetime characteristics under normal operating conditions. The discussions provided here are primarily from the works carried out by Ling et al. (2016).

6.2 Proportional Hazards Models

Suppose there are L distinct inspection times in constant-stress accelerated life-tests (CSALTs), $0 < \tau_1 < \tau_2 < \cdots < \tau_L$. Here, one-shot device testing data are represented as $\mathbf{z} = \{n_{i,l}, K_{i,l}, \mathbf{x}_i, \tau_l\}$, for $i = 1, 2, \ldots, I, l = 1, 2, \ldots, L$, wherein $n_{i,l}$ out of $K_{i,l}$ one-shot devices subject to stress levels $\mathbf{x}_i = (x_{i1}, x_{i2}, \ldots, x_{iJ})$ failed the test at inspection time τ_l. The data are summarized in Table 6.1.

Under the proportional hazards assumption (Cox, 1972; Cox and Oakes, 1984), in the ith test group, the cumulative hazard function at inspection time τ_l is given by

$$H(\tau_l; \nu_i) = H_0(\tau_l)\nu_i, \tag{6.1}$$

Accelerated Life Testing of One-shot Devices: Data Collection and Analysis, First Edition.
Narayanaswamy Balakrishnan, Man Ho Ling, and Hon Yiu So.

Companion Website: www.wiley.com/go/Balakrishnan/Accelerated_Life_Testing

Table 6.1 CSALTs with I test groups and L inspection times for one-shot device testing data analysis under proportional hazards models.

Test group	Stress level	Inspection time τ_1	τ_2	$\cdots$	τ_L
1	$\mathbf{x}_1$	$K_{1,1}$	$K_{1,2}$	$\cdots$	$K_{1,L}$
2	$\mathbf{x}_2$	$K_{2,1}$	$K_{2,2}$	$\cdots$	$K_{2,L}$
$\vdots$	$\vdots$	$\vdots$	$\vdots$	$\ddots$	$\vdots$
I	$\mathbf{x}_I$	$K_{I,1}$	$K_{I,2}$	$\cdots$	$K_{I,L}$

$K_{i,l}$: Number of tested devices in the ith test group at inspection time τ_l.

where $H_0(\tau_l)$ is the baseline cumulative hazard function at inspection time τ_l. We now assume a log-linear link function to relate stress levels $\mathbf{x}_i = (x_{i1}, x_{i2}, \ldots, x_{iJ})$ to lifetime characteristics of devices under normal operating conditions $\mathbf{x}_0 = (x_{01}, x_{02}, \ldots, x_{0J})$ to obtain the cumulative hazard function as in (6.1) with

$$\nu_i = \exp\left(\sum_{j=1}^{J} h_j x_{ij}\right). \tag{6.2}$$

We then obtain the reliability function as

$$R_{\mathrm{PH}}(\tau_l; \nu_i) = \exp(-H(\tau_l; \nu_i)) = (R_0(\tau_l))^{\exp\left(\sum_{j=1}^{J} h_j x_{ij}\right)}, \tag{6.3}$$

where $R_0(\tau_l) = \exp(-H_0(\tau_l))$ is the baseline reliability function at inspection time τ_l. It is of interest to note that the baseline reliability function is a decreasing function of time and bounded between 0 and 1, that is

$$0 \leq R_0(\tau_L) \leq R_0(\tau_{L-1}) \leq \cdots \leq R_0(\tau_2) \leq R_0(\tau_1) \leq 1,$$

and therefore, we let

$$\xi(g_L) = 1 - R_0(\tau_L) = 1 - \exp(-\exp(g_L)), \tag{6.4}$$

and, for $l = 1, 2, \ldots, L-1$,

$$\xi(g_l) = \frac{1 - R_0(\tau_l)}{1 - R_0(\tau_{l+1})} = 1 - \exp(-\exp(g_l)). \tag{6.5}$$

Upon defining $G_l = \prod_{m=l}^{L} \xi(g_m)$, we have

$$R_0(\tau_l) = 1 - \prod_{m=l}^{L}(1 - \exp(-\exp(g_m))) = 1 - G_l. \tag{6.6}$$

Also, it can be seen that $R_0(\tau_l)$ is close to $R_0(\tau_{l+1})$ when g_l tends to $+\infty$, and $R_0(\tau_l)$ is much larger than $R_0(\tau_{l+1})$ when g_l tends to $-\infty$.

The reliability of device at inspection time τ_l under normal operating conditions $\mathbf{x}_0 = (x_{01}, x_{02}, \dots, x_{0J})$ is then given by

$$R_{\text{PH}}(\tau_l) = (R_0(\tau_l))^{\nu_0}, \tag{6.7}$$

where $\nu_0 = \exp\left(\sum_{j=1}^{J} h_j x_{0j}\right)$.

Therefore, the new model parameter of the proportional hazards model is

$$\boldsymbol{\theta}_{\text{PH}} = (g_1, g_2, \dots, g_L, h_1, h_2, \dots, h_J),$$

where g_l and h_j represent the baseline of lifetime distribution at inspection time τ_l and the hazard rate of the jth stress factor, respectively.

We now present a connection between the proportional hazards model and Weibull lifetime distribution that possesses proportional hazard rates. In CSALTs, for $i = 1, 2, \dots, I$, if the lifetimes of devices subject to elevated stress levels $\mathbf{x}_i$ follow Weibull distribution with constant shape parameter over stress levels, that is $\eta_i = \eta = \exp(r_0)$, and scale parameter $\beta_i = \exp\left(\sum_{j=0}^{J} s_j x_{ij}\right)$, with $x_{i0} \equiv 1$, the cdf of the Weibull distribution is given by

$$F_W(t; \beta_i, \eta) = 1 - \exp\left(-\left(\frac{t}{\beta_i}\right)^{\eta}\right), \quad t > 0.$$

If the proportional hazards assumption holds, then the baseline reliability and the coefficients of stress factors are given by

$$R_0(\tau_l) = \exp(-\exp(\eta(\ln(\tau_l) - s_0)))$$

and

$$h_j = -\eta s_j,$$

for $j = 1, 2, \dots, J$.

6.3 Likelihood Inference

The likelihood function, based on the observed data $\mathbf{z}$, is then given by

$$L(\boldsymbol{\theta}_{\text{PH}}) \propto \prod_{i=1}^{I}\prod_{l=1}^{L}(1 - R_{\text{PH}}(\tau_l; \nu_i))^{n_{i,l}}(R_{\text{PH}}(\tau_l; \nu_i))^{K_{i,l}-n_{i,l}}.$$

In the present case, the MLE of model parameter $\boldsymbol{\theta}_{\text{PH}}$ has no explicit form, but can be determined numerically by using standard optimization tools in statistical packages. These optimization tools always require an initial guess of the parameter to obtain the MLE. We now describe a two-stage method for finding initial values for the components of the model parameter. In the first stage, we can write

$$\ln(-\ln(R(\tau_l; \nu_i))) = c_l + \sum_{j=1}^{J} h_j x_{ij},$$

where $c_l = \ln(H_0(\tau_l))$. Then, upon setting $R(\tau_l; \nu_i) = 1 - n_{i,l}/K_{i,l}$, the least-squares method can be employed to estimate c_l and h_j, for $l = 1, 2, \ldots, L$ and $j = 1, 2, \ldots, J$.

In the second stage, we obtain

$$\xi(g_l) = \ln\left(-\ln\left(1 - \frac{1 - R_0(\tau_l)}{1 - R_0(\tau_{l+1})}\right)\right), \tag{6.8}$$

where $R_0(\tau_{L+1}) = 0$ and $R_0(\tau_l) = \exp(-\exp(c_l))$, and consequently the estimate of g_l can also be obtained from (6.8). In addition, the reliability at inspection time τ_l under normal operating conditions $\mathbf{x}_0$ can then be estimated upon substituting $\hat{\boldsymbol{\theta}}_{\text{PH}}$ into the reliability function $R_{\text{PH}}(\tau_l)$ in (6.7).

To construct asymptotic/approximate confidence intervals for the model parameters, the first-order derivatives of reliability for the proportional hazards models with respect to the model parameter are required for the observed Fisher information matrix and are presented below:

$$\begin{aligned}
\frac{\partial R_{\text{PH}}(\tau_l; \nu_i)}{\partial g_m} &= \frac{\partial (R_0(\tau_l))^{\nu_i}}{\partial g_m} = 0, \quad m < l,\\
\frac{\partial R_{\text{PH}}(\tau_l; \nu_i)}{\partial g_m} &= -\nu_i (R_0(\tau_l))^{\nu_i - 1} G_l \left(\frac{(1 - \xi(g_m)) \log(1 - \xi(g_m))}{\xi(g_m)}\right), \quad m \geq l,\\
\frac{\partial R_{\text{PH}}(\tau_l; \nu_i)}{\partial h_j} &= \frac{\partial \exp(-H(\tau_l; \nu_i))}{\partial h_j} = R_{\text{PH}}(\tau_l; \nu_i) H_0(\tau_l) \nu_i x_{ij}.
\end{aligned}$$

Then, the observed Fisher information matrix is given by

$$I_{\text{obs}}(\boldsymbol{\theta}_{\text{PH}}) = \sum_{i=1}^{I} \sum_{l=1}^{L} K_{i,l} \left(\frac{1}{1 - R_{i,l}} + \frac{1}{R_{i,l}}\right) \left(\frac{\partial R_{i,l}}{\partial \boldsymbol{\theta}_{\text{PH}}}\right) \left(\frac{\partial R_{i,l}}{\partial \boldsymbol{\theta}'_{\text{PH}}}\right), \tag{6.9}$$

where $R_{i,l} = R_{\text{PH}}(\tau_l; \nu_i)$. Furthermore, an estimate of the asymptotic variance of the MLE of reliability at inspection time τ_l under normal operating condition $\mathbf{x}_0$ can then be obtained by using delta method as

$$\hat{V}_R = P' \hat{V}_{\theta_{\text{PH}}} P, \tag{6.10}$$

where $P = \partial R_{\text{PH}}(\tau_l) / \partial \boldsymbol{\theta}_{\text{PH}}$ is a column vector of the first-order derivatives, $\nu_0 = \exp\left(\sum_{j=1}^{J} h_j x_{0j}\right)$ and $\hat{V}_{\theta_{\text{PH}}} = I_{\text{obs}}^{-1}(\hat{\boldsymbol{\theta}}_{\text{PH}})$ is the inverse of the observed Fisher information matrix evaluated at the MLE of model parameter, $\hat{\boldsymbol{\theta}}_{\text{PH}}$. In addition, the asymptotic/approximate confidence intervals for reliability can also be constructed using the estimate of the asymptotic variance in (6.10). Details on the derivation of the asymptotic/approximate confidence intervals for reliability have been discussed earlier in Section 2.4.

Hence, the $100(1-\delta)\%$ asymptotic confidence interval for reliability $R_l = R_{\text{PH}}(\tau_l)$ is given by

$$\left(\hat{R}_l - z_{\frac{\delta}{2}}\,\text{se}(\hat{R}_l),\quad \hat{R}_l + z_{\frac{\delta}{2}}\,\text{se}(\hat{R}_l)\right),$$

where z_δ is the upper δth percentile of the standard normal distribution and $\text{se}(\hat{R}_l) = \sqrt{\hat{V}_R}$ is the standard error of $\hat{R}_l = \hat{R}_{\text{PH}}(\tau_l)$. Truncation on the bounds of confidence intervals for reliability may be needed.

Similarly, an approximate $100(1-\delta)\%$ confidence interval based on the hyperbolic arcsecant (arsech) transformation for reliability R_l is given by

$$\left(\frac{2}{\exp(U_f)+\exp(-U_f)},\ \frac{2}{\exp(L_f)+\exp(-L_f)}\right),$$

where $\hat{f} = \ln\left(\frac{1+\sqrt{1-\hat{R}_l^2}}{\hat{R}_l}\right)$, $U_f = \hat{f} + z_{\frac{\delta}{2}}\,\text{se}(\hat{f})$, $L_f = \max\left(0, \hat{f} - z_{\frac{\delta}{2}}\,\text{se}(\hat{f})\right)$ and $\text{se}(\hat{f}) = \frac{\text{se}(\hat{R}_l)}{\hat{R}_l\sqrt{1-\hat{R}_l^2}}$.

Moreover, an approximate $100(1-\delta)\%$ confidence interval based on the logit transformation for reliability R_l is

$$\left(\frac{\hat{R}_l}{\hat{R}_l + (1-\hat{R}_l)S},\ \frac{\hat{R}_l}{\hat{R}_l + (1-\hat{R}_l)S^{-1}}\right),$$

where $S = \exp\left(z_{\frac{\delta}{2}}\,\text{se}(\hat{f})\right)$ and $\text{se}(\hat{f}) = \frac{\text{se}(\hat{R}_l)}{\hat{R}_l(1-\hat{R}_l)}$.

6.4 Test of Proportional Hazard Rates

The proportional hazards models have only the proportional hazards assumption and allow hazard rates to change in a nonparametric way. Here, a test for proportional hazards assumption based on one-shot device testing data is presented. The distance-based test statistics discussed earlier in Chapter 4 can be used to assess the proportional hazards assumption. For example, the distance-based test statistic simply assesses the fit of an assumed distribution to one-shot device testing data, which is given by

$$M = \max_{i,l} |n_{i,l} - K_{i,l}(1 - \hat{R}(\tau_l; v_i))|.$$

When the assumed distribution does not fit the data well, we would observe large M and small p-value, in which case we can readily conclude that the proportional hazards assumption is consonant with the observed data.

6.5 Simulation Studies

In this section, an extensive Monte Carlo simulation study is conducted to examine the performance of the proposed estimation method in terms of bias, root mean square errors, coverage probabilities, and average widths of 95% confidence intervals, based on 10000 Monte Carlo simulations. Suppose the lifetimes of devices follow Weibull distribution with constant shape parameter. All devices were divided into four groups, subject to two stress factors with two elevated stress levels each; that is $(x_{i1}, x_{i2}) \in \{(30, 40), (40, 40), (30, 50), (40, 50)\}$. Then, the devices were inspected at three different times, $(\tau_1, \tau_2, \tau_3) = (10, 20, 30)$. Assuming that $(r_0, s_0, s_1, s_2) = (0.5, 6, -0.03, -0.03)$, the performance of the proposed estimation method for model parameters (h_1, h_2) and reliability $R_{PH}(\tau_l)$ under normal operating conditions $\mathbf{x}_0 = (25, 35)$ is evaluated for different sample sizes $K = K_{i,l} \in \{30, 50, 200\}$, for $i = 1, 2, 3, 4$, and $l = 1, 2, 3$.

Furthermore, the simulated values of the measures for the proportional hazards models are presented in Table 6.2. As expected, the MLEs from the estimation method converge to the true parameters as seen from small values of bias and RMSEs. Moreover, coverage probabilities of the confidence intervals are reasonably close to the nominal level. In addition, the asymptotic and approximate confidence intervals developed here under the proportional hazards models can be compared. Table 6.2 also presents the performance of the three confidence intervals for reliability under the normal operating conditions $\mathbf{x}_0 = (25, 35)$ at different inspection times. It is clear that each estimate tends to the true value accurately, and the coverage probability is close to the nominal level with a larger sample size, resulting in a smaller width. We also observe that the asymptotic confidence intervals are satisfactory for reliability only when the sample size is sufficiently large, and that the approximate confidence intervals have similar behavior and are satisfactory for reliability for all considered cases.

Furthermore, the power of the test for the proportional hazards assumption based on the distance-based test statistic is also evaluated for various sample sizes. The setting of simulation study for this power analysis is similar to the previous simulation study. But, we assume that lifetimes of devices follow Weibull distribution with nonconstant shape parameter $\eta_i = \exp(r_0 + r_1 x_{i1} + r_2 x_{i2})$ with $(r_1, r_2) \in \{(-0.1, 0.1), (-0.11, 0.11), (-0.12, 0.12), (-0.13, 0.13)\}$. Small values of r_1 and r_2 mean that the Weibull distribution possesses hazard rates that are close to being proportional, while large values of r_1 and r_2 mean that the proportional hazards assumption is seriously violated. Furthermore, we consider CSALTs with three elevated stress levels each; that is $x_{i1} \in \{30, 40, 50\}$ and $x_{i2} \in \{40, 50, 60\}$. Here, we consider small sample size $K \in \{13, 22, 88\}$ to maintain the total number of test devices to be similar. Table 6.3 presents the percentage of cases with p-value less than 0.05 for various sample sizes and different values of r_1 and r_2. We observe

Table 6.2 Bias, RMSEs, coverage probabilities, and average widths of 95% confidence intervals for the model parameters and reliability for the proportional hazards model.

	Bias			RMSE		
Parameter	$K = 30$	$K = 50$	$K = 200$	$K = 30$	$K = 50$	$K = 200$
h_1	4.496e−4	2.179e−4	−1.953e−4	1.973e−2	1.500e−2	7.463e−3
h_2	2.327e−4	−3.519e−4	−4.515e−4	1.935e−2	1.528e−2	7.541e−3
$R_{PH}(10)$	−1.375e−3	−1.020e−3	−5.809e−4	1.788e−2	1.359e−2	6.749e−3
$R_{PH}(20)$	−3.883e−3	−3.249e−3	−1.473e−3	4.253e−2	3.321e−2	1.618e−2
$R_{PH}(30)$	−5.351e−3	−4.729e−3	−2.609e−3	6.736e−2	5.241e−2	2.599e−2

	Coverage probability			Average width		
Parameter	$K = 30$	$K = 50$	$K = 200$	$K = 30$	$K = 50$	$K = 200$
h_1	0.948	0.950	0.950	0.076	0.059	0.029
h_2	0.955	0.948	0.950	0.076	0.059	0.029
$R_{PH}(10)$	0.915[a)]	0.930[a)]	0.949[a)]	0.068[a)]	0.053[a)]	0.026[a)]
	0.937[b)]	0.944[b)]	0.949[b)]	0.067[b)]	0.052[b)]	0.026[b)]
	0.952[c)]	0.954[c)]	0.950[c)]	0.074[c)]	0.055[c)]	0.027[c)]
$R_{PH}(20)$	0.930[a)]	0.936[a)]	0.949[a)]	0.163[a)]	0.127[a)]	0.063[a)]
	0.944[b)]	0.946[b)]	0.950[b)]	0.160[b)]	0.125[b)]	0.063[b)]
	0.953[c)]	0.949[c)]	0.948[c)]	0.167[c)]	0.129[c)]	0.063[c)]
$R_{PH}(30)$	0.933[a)]	0.938[a)]	0.952[a)]	0.260[a)]	0.202[a)]	0.101[a)]
	0.947[b)]	0.946[b)]	0.953[b)]	0.252[b)]	0.199[b)]	0.101[b)]
	0.950[c)]	0.949[c)]	0.952[c)]	0.257[c)]	0.201[c)]	0.101[c)]

$h_1 = 0.049$, $h_2 = 0.049$, $R_{PH}(10) = 0.957$, $R_{PH}(20) = 0.872$, $R_{PH}(30) = 0.765$.

a) Asymptotic confidence intervals.

b) Approximate confidence intervals based on arsech transformation.

c) Approximate confidence intervals based on logit transformation.

Table 6.3 Power analysis of distance-based test statistic M for testing the proportional hazards assumption in CSALTs with two and three stress levels.

CSALTs with two stress levels			
(r_1, r_2)	$K = 30$	$K = 50$	$K = 200$
$(-0.1, 0.1)$	0.042	0.123	0.987
$(-0.11, 0.11)$	0.048	0.176	0.998
$(-0.12, 0.12)$	0.056	0.231	0.998
$(-0.13, 0.13)$	0.088	0.276	1.000
CSALTs with three stress levels			
(r_1, r_2)	$K = 13$	$K = 22$	$K = 88$
$(-0.02, 0.02)$	0.043	0.066	0.181
$(-0.03, 0.03)$	0.070	0.114	0.702
$(-0.04, 0.04)$	0.133	0.271	0.980
$(-0.05, 0.05)$	0.238	0.553	1.000

that the power of the test statistic increases with sample size and the values of r_1 and r_2. Moreover, it is worth noting that the number of stress levels is important for testing the proportional hazards assumption based on the distance-based test statistic, as the results reveal that the power of the test statistic for CSALTs with three stress levels is large, compared to CSALTs with two stress levels, even though the values of r_1 and r_2 are small. This may be because more information on hazard rates is gained from CSALTs with more stress levels.

6.6 Case Studies with R Codes

In this section, we consider mice tumor toxicological data in Table 1.5. It is worth noting that the predictors used in this case are x_1 (0 for females, 1 for males) and

Table 6.4 R codes for point estimation and 95% asymptotic/approximate confidence intervals for proportional hazards model.

```
source("PHM.R")
x0<-c(0,sqrt(60))
PHM<-PH(K,n,X,Tau,x0,0.95)
```

$\sqrt{x_2}$ corresponding to the dosage of benzidine dihydrochloride (ppm) used in the study. Furthermore, the R codes for the point estimation as well as for the 95% confidence intervals of the model parameters and the probability of female mice receiving 60 ppm benzidine dihydrochloride not developing tumor by different inspection times for the proportional hazards model as well as for exponential, gamma, and Weibull distributions are displayed in Table 6.4. Finally, the outputs for the model are presented in Tables 6.5 and 6.6. In this case, the p-values of the distance-based test statistic show that the proportional hazards model and exponential distribution do not fit the data well, implying that the hazard rates vary

Table 6.5 R outputs for inference on model parameters based on mice tumor toxicological data in Table 1.5 for the proportional hazards model as well as for exponential, gamma, and Weibull distributions.

Model/Distribution	Parameter	MLE	Asymptotic
Proportional hazards	g_1	−2.024[a]	(−2.545,−1.503)[b]
$M = 7.396$[c]	g_2	−1.091[a]	(−1.523, −0.658)[b]
p-Value = 0.056[d]	g_3	−2.270[a]	(−2.828, −1.713)[b]
	h_1	−2.038[a]	(−2.469, −1.608)[b]
	h_2	0.261[a]	(0.212, −0.311)[b]
Exponential	e_0	−5.457	(−5.954, −4.961)
$M = 15.527$	e_1	−1.204	(−1.539, −0.868)
p-Value = 1.157e-5	e_2	0.173	(0.137, 0.210)
Gamma	a_0	3.119	(2.103, 4.134)
$M = 4.460$	a_1	−1.462	(−2.072, −0.852)
p-Value = 0.462	a_2	0.017	(−0.064, 0.098)
	b_0	0.254	(−0.807, 1.316)
	b_1	2.008	(1.309, 2.706)
	b_2	−0.077	(−0.160, 0.007)
Weibull	r_0	2.413	(1.815, 3.011)
$M = 5.145$	r_1	−0.609	(−0.994, −0.225)
p-Value = 0.285	r_2	−0.043	(−0.093, 0.006)
	s_0	3.382	(3.275, 3.489)
	s_1	0.522	(0.374, 0.670)
	s_2	−0.056	(−0.064, −0.047)

a) PHM$theta.hat.
b) PHM$ACI.
c) PHM$M.
d) PHM$pv.

Table 6.6 R outputs for inference on the probability of female mice ($x_1 = 0$) receiving 60 ppm benzidine dihydrochloride not developing tumor by different inspection times based on mice tumor toxicological data in Table 1.5 for the proportional hazards model as well as for exponential, gamma, and Weibull distributions.

Parameter	MLE	Asymptotic	arsech	logit
$R_{PH}(9.33)$	0.974[a)]	(0.960, 0.988)[b)]	(0.958, 0.986)[c)]	(0.955, 0.985)[d)]
$R_{PH}(14)$	0.807[a)]	(0.750, 0.864)[b)]	(0.749, 0.862)[c)]	(0.744, 0.858)[d)]
$R_{PH}(18.67)$	0.458[a)]	(0.349, 0.568)[b)]	(0.359, 0.577)[c)]	(0.352, 0.568)[d)]
$R_E(9.33)$	0.859	(0.826, 0.892)	(0.825, 0.890)	(0.823, 0.889)
$R_E(14)$	0.796	(0.750, 0.842)	(0.749, 0.840)	(0.746, 0.838)
$R_E(18.67)$	0.737	(0.681, 0.794)	(0.681, 0.793)	(0.677, 0.790)
$R_G(9.33)$	0.999	(0.996, 1.000)	(0.994, 1.000)	(0.987, 1.000)
$R_G(14)$	0.895	(0.837, 0.953)	(0.832, 0.946)	(0.822, 0.940)
$R_G(18.67)$	0.442	(0.332, 0.552)	(0.342, 0.562)	(0.336, 0.553)
$R_W(9.33)$	0.997	(0.992, 1.000)	(0.990, 1.000)	(0.987, 0.999)
$R_W(14)$	0.920	(0.873, 0.968)	(0.868, 0.962)	(0.858, 0.957)
$R_W(18.67)$	0.437	(0.310, 0.564)	(0.324, 0.578)	(0.317, 0.566)

a) PHM$R0;
b) PHM$AsyR;
c) PHM$ArsR;
d) PHM$LogR.

with gender and dosage of chemical, and that the shape parameter is not equal to 1. On the other hand, the gamma and Weibull distributions with varying shape and scale parameters over gender and dosage are suitable for fitting the data. Both distributions agree with that gender has a significant effect on onset time for tumor. In addition, the effect of dosage of benzidine dihydrochloride on the onset time is confirmed under the Weibull distribution. Finally, it is worth noting that the proportional hazards model has its own limitations. One of the limitations is that the estimates of model parameters may not converge when the number of parameters is large. For example, the proportional hazards model may not work well when data involve with many inspection times. Also, due to the very nature of the developed model, it cannot provide any information on reliability after the last inspection time. As a result, the estimation of mean lifetime cannot be obtained.

7

Optimal Design of Tests

7.1 Brief Overview

Optimal design is of great importance in reliability problem as it would facilitate savings in both time and cost. In this chapter, we focus on the design of constant-stress accelerated life-tests (CSALTs) for one-shot devices by assuming Weibull distribution as a lifetime distribution. We then describe a method for determining the optimal allocation of devices, inspection frequency, and the number of inspections at each stress level, subject to a prespecified experimental budget and a termination time. The discussions provided here are primarily from the works carried out by Balakrishnan and Ling (2014b).

7.2 Optimal Design of CSALTs

Consider CSALTs with I test groups, subject to J types of stress factors with higher-than-usual stress levels $\mathbf{x}_i = (x_{i1}, x_{i2}, \ldots, x_{iJ})$. Suppose, in the ith test group, K_i one-shot devices are inspected at L_i equally spaced time points with frequency *freq*. Specifically, $K_{i,l}$ devices are drawn and inspected at a specific time $\tau_{i,l}$, with $\sum_{l=1}^{L_i} K_{i,l} = K_i$. Then, the number of failed devices, $n_{i,l}$, at inspection time $\tau_{i,l}$ are observed, for $l = 1, 2, \ldots, L_i$ and $i = 1, 2, \ldots, I$.

Given a prespecified budget, an optimal test plan that minimizes the asymptotic variance of an estimator of a lifetime parameter of interest will be of interest to determine as it would facilitate efficient collection of data from CSALTs. Because of their practical importance, much work has been done on optimal test plans with cost considerations. Tseng et al. (2009), for example, developed optimal test plans for step-stress accelerated degradation data under a gamma degradation process, and discussed specifically the determination of sample size, measurement frequency, and termination time at each stress level. For this

Accelerated Life Testing of One-shot Devices: Data Collection and Analysis, First Edition.
Narayanaswamy Balakrishnan, Man Ho Ling, and Hon Yiu So.

Companion Website: www.wiley.com/go/Balakrishnan/Accelerated_Life_Testing

purpose, they minimized the asymptotic variance of the maximum likelihood estimator of mean lifetime subject to a prespecified budget. Subsequently, Zhang et al. (2011) considered sample size, stress level, and test time at each stress level as design parameters in the design of an optimal test by minimizing the mean square error of the estimate of reliability at a specific mission time.

To design efficient CSALTs for one-shot devices, we focus on determining the optimal design parameters of inspection frequency, number of inspections at each condition, and number of devices to be allocated. For this purpose, we consider minimizing the asymptotic variance of the maximum likelihood estimator of reliability at a mission time under normal operating conditions, subject to a prespecified budget and a termination time. The required optimization problem would then require the Fisher information matrix for model parameters as well as the first-order derivatives of reliability function with respect to the model parameters; see Chapter 2 for pertinent details. The Fisher information matrix, based on a test plan consisting of inspection frequency, number of inspections at each condition, and allocation of devices, namely, $\zeta = (freq, L_i, K_{i,l})$, is given by

$$I_{\text{obs}}(\boldsymbol{\theta}; \boldsymbol{\zeta}) = \sum_{i=1}^{I} \sum_{l=1}^{L_i} K_{i,l} \left(\frac{1}{F(\tau_{i,l}; \boldsymbol{\theta})} + \frac{1}{R(\tau_{i,l}; \boldsymbol{\theta})} \right) \left(\frac{\partial F(\tau_{i,l}; \boldsymbol{\theta})}{\partial \boldsymbol{\theta}} \right) \left(\frac{\partial F(\tau_{i,l}; \boldsymbol{\theta})}{\partial \boldsymbol{\theta}'} \right), \tag{7.1}$$

where $R(\tau_{i,l}; \boldsymbol{\theta}) = 1 - F(\tau_{i,l}; \boldsymbol{\theta})$ and $\tau_{i,l} = l \times freq$, for $l = 1, 2, \ldots, L_i$, for equi-spaced time points.

Then, the asymptotic variance of the maximum likelihood estimator of reliability at mission time t under normal operating conditions $\mathbf{x}_0$ can be obtained by delta method as (see Chapter 2)

$$V_R(\boldsymbol{\zeta}) = P_R' V_\theta(\boldsymbol{\zeta}) P_R, \tag{7.2}$$

where P_R is a column vector of the first-order derivatives of reliability function at mission time t under normal operating conditions $\mathbf{x}_0$ with respect to model parameters and $V_\theta(\boldsymbol{\zeta})$ is the inverse of the Fisher information matrix $I_{\text{obs}}(\boldsymbol{\theta}; \boldsymbol{\zeta})$ given in (7.1).

7.3 Optimal Design with Budget Constraints

Suppose we have a budget for conducting a CSALT for one-shot devices, the operation cost per unit of time in the ith test group, the cost of devices (including the purchase and testing costs), and the termination time as C_{budget}, $C_{\text{oper},i}$, C_{item}, and τ_{ter}, respectively. It is then evident that a test plan, $\zeta = (freq, L_i, K_{i,l})$, would involve the inspection frequency ($freq$), the number of inspections in the ith test group

($L_i \geq 2$) and the allocation of devices ($K_{i,l}$), for $i = 1, 2, \ldots, I, l = 1, 2, \ldots, L_i$. Then, the total cost of conducting such a CSALT will be

$$TC(\zeta) = C_{\text{item}}\left(\sum_{i=1}^{I}\sum_{l=1}^{L_i} K_{i,l}\right) + freq\left(\sum_{i=1}^{I} C_{\text{oper},i} L_i\right). \tag{7.3}$$

An optimal design will then seek to minimize the asymptotic variance of the maximum likelihood estimator of reliability at mission time t under normal operating conditions $\mathbf{x}_0$ in (7.2), in such a way that the cost considerations are met based on total cost of the experiment given in (7.3), for example.

7.3.1 Subject to Specified Budget and Termination Time

We may be interested in obtaining the optimal CSALTs subject to the constraints $TC(\zeta) \leq C_{\text{budget}}$ and $freq \times L_i \leq \tau_{\text{ter}}$, for all i. Because of the complex form of the objective function in (7.2), we propose the following algorithm for the optimal determination of the design parameters:

Step 1: Given $freq$, find $L_i^* = \left\{\frac{\tau_{\text{ter}}}{freq}\right\}$, for $i = 1, 2, \ldots, I$, where $\lfloor \cdot \rfloor$ is a truncated integer;
Step 2: Set $L_i = 2$, for $i = 1, 2, \ldots, I$;
Step 3: Check if $C_{\text{budget}} \geq C_{\text{item}} K_{\min}\left(\sum_{i=1}^{I} L_i\right) + freq\left(\sum_{i=1}^{I} C_{\text{oper},i} L_i\right)$. If satisfied, determine the optimal allocation of the devices, $K_{i,l}^*$; otherwise, jump to Step 6;
Step 4: Check if $L_i = L_i^*$, for all i. If satisfied, jump to Step 6; otherwise, find the minimum value m such that $L_m < L_m^*$;
Step 5: Check if $m > 1$. If satisfied, set $L_k = 2$, for $k = 1, 2, \ldots, m-1$, and return to Step 3;
Step 6: Set $freq = freq + 1$, and return to Step 1 until $freq = \tau_{\text{ter}}$;
Step 7: The optimal solution of $(freq, L_i, K_{i,l})$ is then given by $min_{(freq, L_i, K_{i,l}^*)}$ $V_R(\zeta(freq, L_i, K_{i,l}^*))$.

In the implementation of the above algorithm, we chose the minimum number of devices allocated at each condition at each inspection time to be $K_{\min} = 20$; an user can change this.

7.3.2 Subject to Standard Deviation and Termination Time

We may instead be interested in minimizing the cost of conducting the experiment subject to a limit on standard deviation of the maximum likelihood estimator of reliability, SD, and termination time. For this purpose, we propose the following algorithm for the optimal determination of the design parameters under this

Table 7.1 R codes for optimal CSALTs subject to a budget of $200 000 and termination time $\tau_{ter} = 36$ for the Weibull distribution.

```
source("CSOTP.R")
CB<-2e5 #budget
CI<-1100 # cost of each device
CO<-c(100,150,200) # costs of operation
X0<-matrix(c(30,40,50),nrow=length(CO),byrow=T)
theta<-c(-0.6,0.03,5.7,-0.05)
x0<-25; t0<-60; Nmin<-20; TT<-36; fq<-3
zeta.B<-BTP(CB,CI,CO,TT,X0,Nmin,x0,t0,theta,model="Weibull",fq)
```

scenario; here again, we chose $K_{\min} = 20$ to be the minimum number of devices allocated at each condition at each inspection time (of course, an user can change this depending on the need):

Step 1: Given *freq*, find $L_i^* = \left\{ \frac{\tau_{ter}}{freq} \right\}$, for $i = 1, 2, \dots, I$;
Step 2: Set $L_i = 2$, for $i = 1, 2, \dots, I$;
Step 3: Find the optimal allocation of devices, $K_{i,l}^*$, such that the asymptotic variance of the maximum likelihood estimator of reliability, $V_R(\zeta(freq, L_i, K_{i,l}^*))$, is less than the square of the specified limit on standard deviation, SD^2, and $\sum_{i=1}^{I} \sum_{l=1}^{L_i} K_{i,l}$ is maximized, and determine $TC(\zeta(freq, L_i, K_{i,l}^*))$;
Step 4: Check if $L_i = L_i^*$, for all i. If satisfied, jump to Step 6; otherwise, find the minimum value m such that $L_m < L_m^*$;
Step 5: Check if $m > 1$. If satisfied, set $L_k = 2$, for $k = 1, 2, \dots, m-1$, and return to Step 3;
Step 6: Set $freq = freq + 1$, and return to Step 1 until $freq = \tau_{ter}$;
Step 7: The optimal solution of $(freq, L_i, K_{i,l})$ is then obtained as $min_{(freq, L_i, K_{i,l}^*)} TC(\zeta(freq, L_i, K_{i,l}^*))$.

The R codes for these two algorithms, for some specifications, are presented in Tables 7.1 and 7.2, for example.

7.4 Case Studies with R Codes

We now illustrate the proposed algorithm of obtaining the optimal CSALTs with a single stress factor under different budgets and different termination times. Let us assume that lifetimes of devices follow the Weibull lifetime distribution with scale

Table 7.2 R codes for optimal CSALTs subject to the limit of standard deviation of 0.1 and termination time $\tau_{ter} = 36$ for the Weibull distribution.

```
source("CSOTP.R")
SD<-0.1 #standard deviation
CI<-1100
CO<-c(100,150,200)
X0<-matrix(c(30,40,50),nrow=length(CO),byrow=T)
theta<-c(-0.6,0.03,5.7,-0.05)
x0<-25; t0<-60; Nmin<-20; TT<-36; fq<-3
zeta.SE<-BTP.se(SD,CI,CO,TT,X0,Nmin,x0,t0,theta,model="Weibull",fq)
```

parameter $\beta_i > 0$ and shape parameter $\eta_i > 0$, wherein the scale and shape parameters are both related to stress levels $\mathbf{x}_i = (x_{i1}, x_{i2}, \dots, x_{iJ})$ through the following log-linear link functions:

$$\eta_i = \exp\left(\sum_{j=0}^{J} r_j x_{ij}\right) \quad \text{and} \quad \beta_i = \exp\left(\sum_{j=0}^{J} s_j x_{ij}\right), \tag{7.4}$$

with $x_{i0} \equiv 1$. The corresponding cdf is

$$F_W(t; \eta_i, \beta_i) = 1 - \exp\left(-\left(\frac{t}{\beta_i}\right)^{\eta_i}\right), \quad t > 0, \tag{7.5}$$

and $\boldsymbol{\theta}_W = \{r_j, s_j, j = 0, 1, \dots, J\}$ is the vector of model parameters; see Chapters 2 for pertinent details.

Suppose the devices have long lives under normal operating conditions, with $(r_0, r_1, s_0, s_1) = (-0.6, 0.03, 5.7, -0.05)$. Now, consider the estimation of reliability at mission time $t = 60$ under a normal operating condition of $\mathbf{x}_0 = 25$, and the CSALTs involve the following setup: $I = 3$ test groups, with minimum number of devices $N_{min} = 20$; termination times $\tau_{ter} \in \{36, 60\}$; elevated stress levels used are $\mathbf{x}_1 = 30, \mathbf{x}_2 = 40$, and $\mathbf{x}_3 = 50$; costs of operation at the elevated stress levels are \$100, \$150, and \$200 per unit of time, respectively; finally, cost of each device is \$1100.

The optimal CSALTs determined, under different budget constraints, by using the algorithm detailed in the last section, are presented in Table 7.3. The corresponding theoretical standard deviation, $s_R = \sqrt{V_R}$, and the empirical standard deviation, $s_{\hat{R}}$ (based on 1000 Monte Carlo simulations), for the estimate of reliability at mission time $t = 60$ were also obtained, and they are presented in the last two columns of Table 7.3. These results show that an experiment with sufficiently large time and/or total budget for the experiment, as one would expect, would result in better reliability estimation.

Table 7.3 Optimal CSALTs with different termination times and under different budgets, along with the corresponding $E[n]$, s_R, and $s_{\hat{R}}$ for reliability at mission time $t = 60$ when the costs of operation at the three stress levels are \$100, \$150, and \$200 per unit of time.

τ_{ter}	C_{budget}	$K_{1,l}$ [a] ($E[n]$)	$K_{2,l}$ [a] ($E[n]$)	$K_{3,l}$ [a] ($E[n]$)	$\tau_{i,l}$ [b]	TC [c]	s_R [d]	$s_{\hat{R}}$
36	\$200 000	20 (3.14)	20 (4.09)	25 (9.33)	18	\$199 900	0.0861	0.1019
		36 (12.70)	46 (25.51)	20 (18.47)	36			
36	\$300 000	20 (3.14)	20 (4.09)	44 (16.42)	18	\$300 000	0.0635	0.0761
		66 (23.29)	88 (48.80)	20 (18.47)	36			
36	\$500 000	20 (3.14)	20 (4.09)	85 (31.71)	18	\$499 100	0.0462	0.0544
		126 (44.45)	168 (93.17)	20 (18.47)	36			
60	\$200 000	20 (4.45)	20 (6.41)	20 (12.25)	24	\$199 800	0.0640	0.0724
		62 (29.36)	20 (14.90)	20 (19.89)	48			
60	\$300 000	20 (2.92)	20 (3.72)	20 (6.67)	17	\$299 700	0.0455	0.0472
		38 (12.60)	20 (10.35)	20 (17.85)	34			
		119 (59.69)			51			
60	\$500 000	20 (2.71)	20 (3.37)	39 (11.50)	16	\$500 000	0.0321	0.0333
		20 (6.20)	20 (9.59)	20 (17.08)	32			
		301 (142.53)			48			

a) zeta.B$K.
b) zeta.B$Tau.
c) zeta.B$Cost.
d) zeta.B$se.

Table 7.4 shows the effect on the optimal CSALTs when the cost of operation at the highest stress level increases from \$200 to \$500. We observe that the results in Tables 7.3 and 7.4 are quite close, except in the total number of test devices. Because the cost of operation is higher in the latter case, the total number of available devices to be tested becomes less, but it seems to have little impact on the asymptotic variance of the maximum likelihood estimator of reliability, V_R. Both theoretical and empirical standard derivations for the estimate of reliability become slightly higher in the second case; moreover, the operation costs may change the inspection times and the number of test groups of CSALTs.

We also determined the optimal CSALTs for different limits on standard deviation and different termination times. The corresponding values of minimum cost and empirical standard deviation (based on 1000 simulated samples) for the estimate of reliability at mission time $t = 60$ were obtained. These results are

Table 7.4 Optimal CSALTs with different termination times and under different budgets, along with the corresponding $E[n]$, s_R, and $s_{\hat{R}}$ for reliability at mission time $t = 60$ when the costs of operation at the three stress levels are \$100, \$150, and \$500 per unit of time.

τ_{ter}	C_{budget}	$K_{1,l}$ [a] ($E[n]$)	$K_{2,l}$ [a] ($E[n]$)	$K_{3,l}$ [a] ($E[n]$)	$\tau_{i,l}$ [b]	TC [c]	s_R [d]	$s_{\hat{R}}$
36	\$200 000	20 (3.14)	20 (4.09)	22 (10.45)	18	\$199 700	0.0903	0.1081
		32 (11.29)	43 (23.85)	20 (18.47)	36			
36	\$300 000	20 (3.14)	20 (4.09)	43 (17.91)	18	\$299 800	0.0651	0.0781
		63 (22.23)	82 (45.48)	20 (18.47)	36			
36	\$500 000	20 (3.14)	20 (4.09)	83 (30.97)	18	\$500 000	0.0468	0.0564
		124 (43.75)	163 (90.40)	20 (18.47)	36			
60	\$200 000	20 (4.45)	20 (6.41)	20 (12.25)	24	\$199 900	0.0689	0.0795
		49 (23.20)	20 (14.90)	20 (19.89)	48			
60	\$300 000	20 (2.92)	20 (3.72)	20 (17.85)	17	\$300 000	0.0466	0.0476
		35 (11.60)	20 (10.35)	20 (17.85)	34			
		113 (56.68)			51			
60	\$500 000	20 (1.49)	20 (1.51)	20 (2.08)	10	\$499 800	0.0324	0.0432
		20 (3.57)	20 (4.84)	40 (18.16)	20			
		20 (5.77)			30			
		20 (7.89)			40			
		258 (127.02)			50			

a) zeta.B$K.
b) zeta.B$Tau.
c) zeta.B$Cost.
d) zeta.B$se.

presented in Table 7.5, and they reveal that the termination time of the experiment has a significant impact on the total cost of conducting the experiment. When the operation cost is relatively low compared to the cost of devices, as one would expect, an experiment with a reasonable time duration would result in reducing the total number of test devices as well as the total cost of conducting the experiment.

All the simulation results show that the termination time is a critical factor in the design of CSALTs for one-shot devices. In typical CSALTs (see Nelson 2009; Meeker and Escobar 2014), the fraction failing at normal operating conditions is usually extremely small. Though the use of higher-than-normal stress levels would induce failures quickly in an accelerated life-test (ALT), as pointed out by Escobar and Meeker (1995), the actual variance may be much larger than

Table 7.5 Optimal CSALTs with different standard deviations, *SD*, of the reliability at mission time $t = 60$ and termination times, along with the corresponding $E[n]$, minimum total cost of conducting the experiment, and s_R when the costs of operation at the three stress levels are \$100, \$150, and \$200 per unit of time.

τ_{ter}	*SD*	$K_{1,l}$[a] ($E[n]$)	$K_{2,l}$[a] ($E[n]$)	$K_{3,l}$[a] ($E[n]$)	$\tau_{i,l}$[b]	*TC*[c]	s_R[d]
36	0.10	20 (3.14)	20 (4.09)	20 (7.46)	18	\$170 200	0.0994
		25 (8.82)	35 (19.41)	20 (18.47)	36		
36	0.08	20 (3.14)	20 (4.09)	29 (10.82)	18	\$218 600	0.0800
		41 (14.46)	54 (29.95)	20 (18.47)	36		
36	0.05	20 (3.14)	20 (4.09)	73 (27.23)	18	\$437 500	0.0500
		107 (37.75)	143 (79.31)	20 (18.47)	36		
60	0.10	20 (3.79)	20 (5.23)	20 (9.89)	21	\$150 900	0.0966
		20 (8.30)	20 (13.15)	20 (19.53)	42		
60	0.08	20 (4.67)	20 (6.81)	20 (12.98)	25	\$164 400	0.0795
		29 (14.28)	20 (15.41)	20 (19.94)	50		
60	0.05	20 (2.92)	20 (3.72)	20 (6.67)	17	\$265 600	0.0499
		30 (9.95)	20 (10.35)	20 (17.85)	34		
		96 (48.15)			51		

a) zeta.SE$K.
b) zeta.SE$Tau.
c) zeta.SE$Cost.
d) zeta.SE$se.

the asymptotic approximation in the optimal test plan, due to the fact that few failures occur at low-stress levels. We did not encounter this problem when N_{min} and τ_{ter} were sufficiently large in the scenario considered here. A test with a longer duration would not only produce more failures but also would effectively control experimental budgets.

Nelson (2009) observed that traditional test plans with equal numbers of devices at equally spaced test stress levels may not be optimal under a right-censoring scheme with a single stress factor, and also listed some references on optimal test plans for inspection data. Meeker and Hahn (1985) introduced 4: 2: 1 design as an alternative to the traditional test plan in the case of three stress factors. Escobar and Meeker (1995) presented guidelines for planning ALTs with two or more stress factors, by finding the optimal test plan for a single stress factor test situation and then systematically splitting the plan's stress level combination into two stress level combination based on D-optimality. It is true that in their planning, the actual variance turns out to be much larger than the asymptotic approximation because of very few failures being observed at some stress levels.

The optimal design method in the last section can also be adopted for CSALTs with two or more stress factors subject to budget and termination time constraints. The objective of CSALTs with several stress factors is naturally to induce more failures at slightly elevated stress levels, through the combined effect of stress factors. Specifically, let lifetimes of devices, subject to elevated stress levels $\mathbf{x}_i = (x_{i1}, x_{i2})$, have Weibull distribution with scale parameter $\beta_i = \exp(r_0 + r_1 x_{i1} + r_2 x_{i2})$ and shape parameter $\eta_i = \exp(s_0 + s_1 x_{i1} + s_2 x_{i2})$. Suppose that the true model parameters are $(r_0, r_1, r_2, s_0, s_1, s_2) = (-0.63, 0.03, 0.0005, 5.76, -0.05, -0.001)$. Now, let us consider the estimation of reliability at mission time $t = 60$ under normal operating conditions of $\mathbf{x}_0 = (x_{01}, x_{02}) = (25, 60)$, and the CSALTs involve the following set up: termination time $\tau_{\text{ter}} \in \{36, 60\}$; elevated stress levels used are $(x_{i1}, x_{i2}) \in \{(30, 70), (40, 70), (50, 80)\}$; costs of operation at the elevated stress levels are \$110, \$160, and \$220 per unit of time, respectively; finally, cost of each device is \$1100.

Then, the optimal CSALTs determined, under different budget constraints, are presented in Table 7.6. From these results, we observe that the empirical standard deviation for the estimate of reliability is much higher than the corresponding theoretical standard deviation, especially in the case of shorter experimental time. For this reason, the use of the asymptotic variance of the maximum likelihood estimator of reliability may not be a reasonable criterion for the optimal design of CSALTs with two or more stress factors. Finally, it is worth noting that the algorithms developed in the preceding section can also be adopted for exponential and gamma lifetime distributions to obtain optimal CSALTs for one-shot devices.

7.5 Sensitivity of Optimal Designs

The effects of mis-specification of model parameters on optimal CSALTs need to be studied for evaluating their robustness feature. As the presumed value $\theta_W^* = (r_0^*, r_1^*, s_0^*, s_0^*)$ is likely to differ from the true model parameter $\theta_W = (r_0, r_1, s_0, s_1)$, we assume here the form $\theta_W^* = (r_0^*, r_1^*, s_0^*, s_0^*) = ((1+\varepsilon_1)r_0, (1+\varepsilon_2)r_1, (1+\varepsilon_3)s_0, (1+\varepsilon_4)s_1)$, where $\varepsilon_i \in \{-0.1, 0.0, 0.1\}$, thus allowing for under-specification as well as over-specification from the true value of $\boldsymbol{\theta}_W$.

With the specification of a total budget of \$200 000 and termination time of 36 months, and under the same set-up as in the preceding section, we determined the optimal CSALTs under various combinations of errors ($\varepsilon_i, i = 1, 2, 3, 4$). These results are all presented in Tables 7.7–7.9, along with variance efficiency (VE) of

Table 7.6 Optimal CSALTs with two stress factors for one-shot devices under different budgets and termination times, along with the corresponding $E[n]$, s_R, and $s_{\hat{R}}$ for reliability at mission time $t = 60$.

τ_{ter}	C_{budget}	$K_{1,l}(E[n])$	$K_{2,l}(E[n])$	$K_{3,l}(E[n])$	$\tau_{i,l}$	TC	s_R	$s_{\hat{R}}$
36	\$200 000	20 (3.15)	25 (5.16)	32 (12.34)	18	\$199 140	0.1086	0.1599
		36 (12.80)	32 (17.95)	20 (18.69)	36			
36	\$300 000	20 (1.88)	20 (2.08)	36 (5.87)	12	\$299 900	0.0830	0.1352
		20 (4.48)	62 (20.09)	21 (13.24)	24			
		51 (18.13)	29 (16.26)		36			
36	\$500 000	34 (3.20)	20 (2.08)	65 (10.59)	12	\$499 000	0.0616	0.0842
		20 (4.48)	114 (36.94)	37 (23.32)	24			
		95 (33.77)	55 (30.84)		36			
60	\$200 000	20 (2.94)	21 (3.95)	29 (9.99)	17	\$198 930	0.0970	0.1340
		20 (6.68)	34 (17.80)	20 (18.12)	34			
		20 (10.11)			51			
60	\$300 000	20 (2.51)	20 (3.05)	49 (13.05)	15	\$299 600	0.0697	0.1032
		20 (5.81)	84 (37.40)	20 (16.46)	30			
		20 (8.96)			45			
		23 (13.44)			60			
60	\$500 000	20 (2.51)	29 (4.43)	98 (26.09)	15	\$499 800	0.0511	0.0731
		38 (11.03)	167 (74.36)	20 (16.46)	30			
		20 (8.96)			45			
		46 (26.89)			60			

the estimator of reliability, at mission time t_0, with the optimal designs based on $\boldsymbol{\theta}_W$ and $\boldsymbol{\theta}_W^*$, computed as

$$VE(R) = \frac{V_R(\zeta_{\theta_W})}{V_R(\zeta_{\theta_W^*})}.$$

The VE measures the goodness of the design under the mis-specified parameter values relative to the optimal design under the true parameter values; see Montgomery (2013), for example. If the VE is close to 1, it shows that the optimal plan is robust. The VE of 0.8 indicates that the optimal plan is sensitive to parameter values as it loses 20% efficiency when the parameter is mis-specified.

Table 7.7 Sensitivity analysis of optimal CSALTs under various combinations of parameters $(0.9r_0, (1+\varepsilon_2)r_1, (1+\varepsilon_3)s_0, (1+\varepsilon_4)s_1)$, with ε_i being departures from the true value of the parameter.

ε_1 (%)	ε_2 (%)	ε_3 (%)	ε_4 (%)	$K_{1,l}$	$K_{2,l}$	$K_{3,l}$	*freq*	*VE*(*R*)
0	**0**	**0**	**0**	**(20,36)**	**(20,46)**	**(25,20)**	**18**	**1.0000**
−10	−10	−10	−10	(20,25)	(20,60)	(22,20)	18	0.9826
−10	−10	−10	0	(20,23)	(20,53)	(31,20)	18	0.9843
−10	−10	−10	+10	(20,35)	(20,44)	(28,20)	18	0.9924
−10	−10	0	−10	(20,24)	(20,55)	(28,20)	18	0.9912
−10	−10	0	0	(20,37)	(20,46)	(24,20)	18	1.0004
−10	−10	0	+10	(20,43)	(20,40)	(24,20)	18	0.9872
−10	−10	+10	−10	(20,21)	(20,53)	(33,20)	18	0.9749
−10	−10	+10	0	(20,32)	(20,46)	(29,20)	18	0.9929
−10	−10	+10	+10	(20,42)	(20,38)	(27,20)	18	0.9764
−10	0	−10	−10	(20,24)	(20,56)	(22,25)	18	0.9447
−10	0	−10	0	(20,20)	(21,53)	(33,20)	18	0.9638
−10	0	−10	+10	(20,57)	(20,29)	(21,20)	18	0.9289
−10	0	0	−10	(20,25)	(20,55)	(27,20)	18	0.9935
−10	0	0	0	(20,36)	(20,44)	(27,20)	18	0.9944
−10	0	0	+10	(20,44)	(20,40)	(23,20)	18	0.9872
−10	0	+10	−10	(20,22)	(20,52)	(33,20)	18	0.9765
−10	0	+10	0	(20,32)	(20,44)	(31,20)	18	0.9834
−10	0	+10	+10	(20,42)	(20,38)	(27,20)	18	0.9764
−10	+10	−10	−10	(20,20,20)	(20,20,29)	(20,20)	12	0.7586
−10	+10	−10	0	(20,20)	(21,55)	(31,20)	18	0.9692
−10	+10	−10	+10	(20,53)	(20,29)	(25,20)	18	0.9265
−10	+10	0	−10	(20,25)	(20,54)	(28,20)	18	0.9930
−10	+10	0	0	(20,37)	(20,45)	(25,20)	18	0.9987
−10	+10	0	+10	(20,46)	(20,39)	(22,20)	18	0.9832
−10	+10	+10	−10	(20,21)	(20,50)	(36,20)	18	0.9618
−10	+10	+10	0	(20,37)	(20,41)	(29,20)	18	0.9825
−10	+10	+10	+10	(20,43)	(20,36)	(28,20)	18	0.9652

Table 7.8 Sensitivity analysis of optimal CSALTs under various combinations of parameters $(r_0, (1+\varepsilon_2)r_1, (1+\varepsilon_3)s_0, (1+\varepsilon_4)s_1)$, with ε_i being departures from the true value of the parameter.

ε_1 (%)	ε_2 (%)	ε_3 (%)	ε_4 (%)	$K_{1,l}$	$K_{2,l}$	$K_{3,l}$	*freq*	*VE*(*R*)
0	**0**	**0**	**0**	**(20,36)**	**(20,46)**	**(25,20)**	**18**	**1.0000**
0	−10	−10	−10	(20,24)	(20,60)	(23,20)	18	0.9835
0	−10	−10	0	(20,52)	(20,35)	(20,20)	18	0.9639
0	−10	−10	+10	(20,35)	(20,44)	(28,20)	18	0.9924
0	−10	0	−10	(20,24)	(20,56)	(27,20)	18	0.9914
0	−10	0	0	(20,34)	(20,48)	(25,20)	18	1.0016
0	−10	0	+10	(20,43)	(20,41)	(23,20)	18	0.9902
0	−10	+10	−10	(20,22)	(20,54)	(31,20)	18	0.9825
0	−10	+10	0	(20,32)	(20,46)	(29,20)	18	0.9929
0	−10	+10	+10	(20,41)	(20,39)	(27,20)	18	0.9804
0	0	−10	−10	(20,25)	(20,59)	(21,22)	18	0.9644
0	0	−10	0	(20,20)	(21,53)	(33,20)	18	0.9638
0	0	−10	+10	(20,35)	(20,44)	(28,20)	18	0.9924
0	0	0	−10	(20,25)	(20,55)	(27,20)	18	0.9935
0	0	0	+10	(20,44)	(20,40)	(23,20)	18	0.9872
0	0	+10	−10	(20,21)	(20,52)	(34,20)	18	0.9712
0	0	+10	0	(20,32)	(20,45)	(30,20)	18	0.9884
0	0	+10	+10	(20,42)	(20,38)	(27,20)	18	0.9764
0	+10	−10	−10	(20,23)	(20,50)	(20,34)	18	0.8967
0	+10	−10	0	(20,20)	(21,53)	(33,20)	18	0.9638
0	+10	−10	+10	(20,55)	(20,29)	(23,20)	18	0.9286
0	+10	0	−10	(20,25)	(20,54)	(28,20)	18	0.9930
0	+10	0	0	(20,39)	(20,44)	(24,20)	18	0.9974
0	+10	0	+10	(20,46)	(20,39)	(22,20)	18	0.9832
0	+10	+10	−10	(20,21)	(20,53)	(33,20)	18	0.9749
0	+10	+10	0	(20,32)	(20,44)	(31,20)	18	0.9834
0	+10	+10	+10	(20,42)	(20,37)	(28,20)	18	0.9699

Table 7.9 Sensitivity analysis of optimal CSALTs under various combinations of parameters $(1.1r_0, (1+\varepsilon_2)r_1, (1+\varepsilon_3)s_0, (1+\varepsilon_4)s_1)$, with ε_i being departures from the true value of the parameter.

ε_1 (%)	ε_2 (%)	ε_3 (%)	ε_4 (%)	$K_{1,l}$	$K_{2,l}$	$K_{3,l}$	*freq*	*VE*(*R*)
0	**0**	**0**	**0**	**(20,36)**	**(20,46)**	**(25,20)**	**18**	**1.0000**
+10	−10	−10	−10	(20,24)	(20,60)	(23,20)	18	0.9835
+10	−10	−10	0	(20,32)	(20,52)	(23,20)	18	1.0008
+10	−10	−10	+10	(20,32)	(20,46)	(29,20)	18	0.9929
+10	−10	0	−10	(20,24)	(20,56)	(27,20)	18	0.9914
+10	−10	0	0	(20,34)	(20,48)	(25,20)	18	1.0016
+10	−10	0	+10	(20,43)	(20,41)	(23,20)	18	0.9902
+10	−10	+10	−10	(20,21)	(20,54)	(32,20)	18	0.9780
+10	−10	+10	0	(20,37)	(20,45)	(25,20)	18	0.9987
+10	−10	+10	+10	(20,41)	(20,39)	(27,20)	18	0.9804
+10	0	−10	−10	(20,25)	(20,60)	(22,20)	18	0.8926
+10	0	−10	0	(20,20)	(21,53)	(33,20)	18	0.9638
+10	0	−10	+10	(20,35)	(20,44)	(28,20)	18	0.9924
+10	0	0	−10	(20,25)	(20,55)	(27,20)	18	0.9935
+10	0	0	0	(20,35)	(20,48)	(24,20)	18	1.0020
+10	0	0	+10	(20,44)	(20,40)	(23,20)	18	0.9872
+10	0	+10	−10	(20,21)	(20,54)	(32,20)	18	0.9780
+10	0	+10	0	(20,32)	(20,46)	(29,20)	18	0.9929
+10	0	+10	+10	(20,41)	(20,40)	(26,20)	18	0.9855
+10	+10	−10	−10	(20,25)	(20,55)	(20,27)	18	0.9251
+10	+10	−10	0	(20,20)	(20,54)	(32,20)	18	0.9667
+10	+10	−10	+10	(20,33)	(20,44)	(30,20)	18	0.9869
+10	+10	0	−10	(20,25)	(20,54)	(28,20)	18	0.9930
+10	+10	0	0	(20,36)	(20,46)	(25,20)	18	1.0000
+10	+10	0	+10	(20,45)	(20,39)	(23,20)	18	0.9837
+10	+10	+10	−10	(20,21)	(20,52)	(34,20)	18	0.9712
+10	+10	+10	0	(20,32)	(20,44)	(31,20)	18	0.9834
+10	+10	+10	+10	(20,43)	(20,37)	(27,20)	18	0.9720

These values are presented in the last column of Tables 7.7–7.9, and they help us understand how the variations in the presumed values of model parameters affect the optimal CSALTs and also the corresponding estimates of reliability based on such a plan. Most importantly, we observe from Tables 7.7–7.9 that the optimal CSALTs under mis-specification of the parameters are quite close to the optimal CSALTs determined under the true parameter specification, thus revealing the natural robustness of the optimal CSALTs developed in this chapter.

8

Design of Simple Step-Stress Accelerated Life-Tests

8.1 Brief Overview

Up to this point, we have provided extensive discussions on inferential methods based on one-shot device testing data obtained from constant-stress accelerated life-tests (CSALTs) as well as the construction of optimal test plans. However, as compared to CSALTs, step-stress accelerated life-tests (SSALTs) are known to be more efficient and would require fewer test units. In this chapter, we therefore provide a detailed discussion on the optimal designs of SSALTs for one-shot devices with exponential and Weibull lifetime distributions. These works were originally developed by Ling (2019) and Ling and Hu (2020).

8.2 One-Shot Device Testing Data Under Simple SSALTs

Consider a simple SSALT wherein the stress level is changed only once during the test. Assume that $0 < \tau_1 < \tau_2$, $0 < K_1 < K$, and $x_1 < x_2$, and that all K devices are exposed to the same initial stress level of x_1. Out of the K devices, K_1 are selected for inspection at a prespecified inspection time τ_1, of which n_1 are recorded to have failed. At this point, the stress level is increased to x_2, and all the remaining $K_2 = K - K_1$ devices are then inspected at another prespecified inspection time τ_2. The number of failures n_2 is then recorded. The one-shot device testing data thus observed can be summarized as in Table 8.1. Let us denote the one-shot device testing data so obtained by $\mathbf{z} = \{\tau_i, K_i, n_i, x_i, i = 1, 2\}$.

Sedyakin (1966) and Nelson (1980) discussed cumulative exposure (CE) models for SSALTs. The CE models assume that the remaining lifetime of a device

Accelerated Life Testing of One-shot Devices: Data Collection and Analysis, First Edition.
Narayanaswamy Balakrishnan, Man Ho Ling, and Hon Yiu So.

Companion Website: www.wiley.com/go/Balakrishnan/Accelerated_Life_Testing

Table 8.1 One-shot device testing data under SSALTs with two stress levels.

Stage	Inspection time	Number of tested devices	# of failures	Stress level
1	τ_1	K_1	n_1	x_1
2	τ_2	K_2	n_2	x_2

depends on the cumulative stress that the device has been exposed to, regardless of how the stress levels were accumulated. Specifically, the surviving devices will fail according to the lifetime distribution at the current stress that accounts for the accumulated stress up to that point in its functional form. The CE model is widely used in reliability studies. Miller and Nelson (1983) and Alhadeed and Yang (2005) discussed optimal simple SSALT plans with the CE model under exponential and log-normal distributions, respectively. Bai et al. (1989) extended the results of Miller and Nelson (1983) to censoring schemes.

Let T denote the lifetime of a device having a distribution with cumulative hazard functions $G_1(t;\theta_1)$ and $G_2(t;\theta_2)$ at stress levels x_1 and x_2, respectively. Under the CE model, the cumulative hazard function is given by

$$H(t) = \begin{cases} G_1(t;\theta_1), & 0 < t \leq \tau_1, \\ G_2(t+(s-1)\tau_1;\theta_2), & t > \tau_1, \end{cases} \tag{8.1}$$

where s is such that

$$G_1(\tau_1;\theta_1) = G_2(s\tau_1;\theta_2). \tag{8.2}$$

The corresponding cumulative distribution function (cdf) is then given by

$$F(t) = 1 - \exp(-H(t)). \tag{8.3}$$

The observed likelihood function, based on (8.3), is then given by

$$L(\theta;\mathbf{z}) \propto \prod_{i=1}^{2} [F(\tau_i;\boldsymbol{\Psi}_i)]^{n_i} [R(\tau_i;\boldsymbol{\Psi}_i)]^{K_i-n_i}, \tag{8.4}$$

where $R(\tau_i;\boldsymbol{\Psi}_i) = 1 - F(\tau_i;\boldsymbol{\Psi}_i)$ and $\boldsymbol{\Psi}_i = \boldsymbol{\Psi}(x_i;\theta)$.

As in Chapter 7, we shall use the Fisher information matrix for the optimal design of SSALTs for one-shot devices. From (8.4), the Fisher information matrix is obtained as

$$I_{\text{obs}}(\theta) = \sum_{i=1}^{2} K_i \left(\frac{1}{F(\tau_i;\theta)} + \frac{1}{R(\tau_i;\theta)} \right) \left(\frac{\partial F(\tau_i;\theta)}{\partial \theta} \right) \left(\frac{\partial F(\tau_i;\theta)}{\partial \theta'} \right). \tag{8.5}$$

For convenience, let us denote $K_1 = K\pi$, $K_2 = K(1-\pi)$, $A_i = \frac{1}{F(\tau_i;\theta)} + \frac{1}{R(\tau_i;\theta)}$ and $X_{ik} = -\partial F(\tau_i;\theta)/\partial\theta_k$,, where θ_k is the kth component of $\boldsymbol{\theta}$.

With this notation, the (p, q)th element of the Fisher information matrix, $I_{\text{obs}}(\boldsymbol{\theta})$ in (8.5), is given by

$$\begin{aligned}&K(\pi A_1X_{1p}X_{1q} + (1-\pi)A_2X_{2p}X_{2q})\\&= K(A_2X_{2p}X_{2q} + \pi(A_1X_{1p}X_{1q} - A_2X_{2p}X_{2q}))\\&= K(r_{pq} + s_{pq}\pi).\end{aligned}$$

8.3 Asymptotic Variance

8.3.1 Exponential Distribution

As in Chapters 2, let us use E to denote the exponential distribution. In this case, we have $G_i(\tau_i; \lambda_i) = \lambda_i t$ for $i = 1, 2$, where λ_1 and λ_2 are the rate parameters at stages 1 and 2 of the SSALT, respectively, while $\beta_i = 1/\lambda_i$ $(i = 1, 2)$ are the mean (scale) parameters. Then, the CE model stipulates, through the solution of (8.2), the corresponding cumulative hazard function, reliability function, and probability density function (pdf) to be

$$H(t) = \begin{cases} \lambda_1 t, & 0 < t \le \tau_1, \\ \lambda_1\tau_1 + \lambda_2(t-\tau_1), & t > \tau_1, \end{cases} \tag{8.6}$$

$$R(t) = \exp(-H(t)) = \begin{cases} \exp(-\lambda_1 t), & 0 < t \le \tau_1, \\ \exp\{-(\lambda_1\tau_1 + \lambda_2(t-\tau_1))\}, & t > \tau_1, \end{cases} \tag{8.7}$$

and

$$f(t) = -R'(t) = \begin{cases} \lambda_1 \exp(-\lambda_1 t), & 0 < t \le \tau_1, \\ \lambda_2 \exp\{-(\lambda_1\tau_1 + \lambda_2(t-\tau_1))\}, & t > \tau_1. \end{cases} \tag{8.8}$$

We further assume that the rate parameters are related to the stress levels in a log-linear form as

$$\lambda_i = \exp(e_0 + e_1 x_i), \tag{8.9}$$

and let $\boldsymbol{\theta}_E = (e_0, e_1)$ denote the new model parameter. Observe, from (8.9), that when the slope parameter $e_1 > 0$, then the effect of increasing the stress level from x_1 to x_2 will be in increasing the rate parameter from λ_1 to λ_2 or, equivalently, decreasing the mean lifetime from β_1 to β_2. Conversely, when $e_1 < 0$, the effect will be to decrease the rate parameter or, equivalently, to increase the mean lifetime.

Therefore, we have

$$\begin{aligned}X_{ik} &= -\frac{\partial F(\tau_i; \boldsymbol{\theta}_E)}{\partial e_k} = d_{ik}R(\tau_i; \boldsymbol{\theta}_E),\\ d_{1k} &= \lambda_1\tau_1 x_1^k,\\ d_{2k} &= \lambda_1\tau_1 x_1^k + \lambda_2(\tau_2 - \tau_1)x_2^k.\end{aligned}$$

Then, the Fisher information matrix in (8.5) can be shown to be

$$I_{\text{obs}}(\theta) = K \begin{bmatrix} r_{00} + s_{00}\pi & r_{10} + s_{10}\pi \\ r_{10} + s_{10}\pi & r_{11} + s_{11}\pi \end{bmatrix}, \tag{8.10}$$

where $r_{kk} = A_2X_{2k}^2, r_{10} = A_2X_{21}X_{20}, s_{kk} = A_1X_{1k}^2 - A_2X_{2k}^2$, and $s_{10} = A_1X_{11}X_{10} - A_2X_{21}X_{20}$. The asymptotic variance–covariance matrix of the maximum likelihood estimator (MLE) of model parameter $\theta_E = (e_0, e_1)$ is then obtained readily from (8.10) as

$$V_{\theta_E} = (I_{\text{obs}}(\theta_E))^{-1} = \frac{1}{D} \begin{bmatrix} r_{11} + s_{11}\pi & -(r_{10} + s_{10}\pi) \\ -(r_{10} + s_{10}\pi) & r_{00} + s_{00}\pi \end{bmatrix}, \tag{8.11}$$

where $D = K\{(r_{00} + s_{00}p_1)(r_{11} + s_{11}p_1) - (r_{10} + s_{10}\pi)^2\}$.

By using delta method, the asymptotic variance of the MLE of mean lifetime under the normal operating condition x_0, for example, becomes

$$V_\mu = P'_\mu V_{\theta_E} P_\mu, \tag{8.12}$$

where

$$P_\mu = \begin{bmatrix} \frac{\partial \mu}{\partial e_0} \\ \frac{\partial \mu}{\partial e_1} \end{bmatrix} = \begin{bmatrix} -\frac{1}{\lambda_0} \\ -\frac{x_0}{\lambda_0} \end{bmatrix}.$$

Thus, (8.12) yields

$$\begin{aligned} V_\mu &= \frac{(r_{11} - 2r_{10}x_0 + r_{00}x_0^2) + (s_{11} - 2s_{10}x_0 + s_{00}x_0^2)\pi}{K\lambda_0^2(r_{00}r_{11} - r_{10}^2) + (r_{00}s_{11} + r_{11}s_{00} - 2r_{10}s_{10})\pi + (s_{00}s_{11} - s_{10}^2)\pi^2} \\ &= \frac{(r_{11} - 2r_{10}x_0 + r_{00}x_0^2) + (s_{11} - 2s_{10}x_0 + s_{00}x_0^2)\pi}{K\lambda_0^2(r_{00}s_{11} + r_{11}s_{00} - 2r_{10}s_{10})\pi(1 - \pi)}. \end{aligned} \tag{8.13}$$

8.3.2 Weibull Distribution

Zheng et al. (2018) considered a SSALT with four stress levels (333, 339, 345, 351 K) to collect grease-based magnetorheological fluids (G-MRFs) samples for predicting the lifetime of G-MRF at the normal operating temperature of 293 K under Weibull distribution. Twenty G-MRF samples were used in the SSALT. At the end of each stress level, five samples were extracted, and their viscosity and shear stress were measured. The extracted samples were removed from the test and so the rheological properties of each sample could be measured only once. The failure of G-MRF occurs when its viscosity or shear stress decreases by more than 10% after each stress level. Their data, collected under a SSALT, have been presented earlier in Table 1.4, which is in the form of one-shot device testing data. Motivated by these data on G-MRF samples collected from a SSALT, we discuss here the determination of design of simple SSALTs for one-shot devices under Weibull lifetime distribution. Note, however, that the data in Table 1.4 is not of the form of a simple SSALT as it involves four stages of test. So the determination of an optimal SSALT for this general case, extending the work presented here for the simple SSALT case, remains as an open problem and will be of great interest to resolve.

At both stages of the simple SSALT, it is assumed that lifetimes of devices follow Weibull distribution with $G_i(\tau_i; \beta_i) = (t/\beta_i)^\eta$ for $i = 1, 2$. Then, under the CE model, proceeding as before with (8.1) and (8.2), we have $s = \beta_2/\beta_1$ and the corresponding cumulative hazard function and reliability function are given by

$$H(t) = \begin{cases} \left(\frac{t}{\beta_1}\right)^\eta, & 0 < t \le \tau_1, \\ \left(\frac{\tau_1}{\beta_1} + \frac{t-\tau_1}{\beta_2}\right)^\eta, & t > \tau_1, \end{cases} \tag{8.14}$$

and

$$R(t) = \exp(-H(t)) = \begin{cases} \exp\left(-\left(\frac{t}{\beta_1}\right)^\eta\right), & 0 < t \le \tau_1, \\ \exp\left(-\left(\frac{\tau_1}{\beta_1} + \frac{t-\tau_1}{\beta_2}\right)^\eta\right), & t > \tau_1, \end{cases} \tag{8.15}$$

where $\beta_1 > 0$ and $\beta_2 > 0$ are scale parameters at the first and second stages, respectively, and η is the common shape parameter. We further assume that the scale parameters relate to stress levels in a log-linear link form as

$$\beta_i = \exp(s_0 + s_1 x_i), \quad i = 0, 1, 2,$$

thus facilitating an extrapolation for the normal operating condition x_0. Then, the reliability at mission time t, under normal operating condition x_0, is given by

$$R(t) = \exp\left\{-\left(\frac{t}{\beta_0}\right)^\eta\right\} = \exp[-\exp\{-\eta(s_0 + s_1 x_0 - \ln(t))\}]. \tag{8.16}$$

Now, $\boldsymbol{\theta}_W = (w_0, w_1, w_2) = (s_0, s_1, \eta)$ becomes the new parameter of this model.

As before, for $i = 1, 2$ and $k = 0, 1$, let us set $K_1 = K\pi, K_2 = K(1-\pi)$ and

$$\begin{aligned} X_{ik} &= -\frac{\partial F(\tau_i; \boldsymbol{\theta}_W)}{\partial w_k} = d_{ik} R(\tau_i; \boldsymbol{\theta}_W) w_2, \\ X_{i2} &= -\frac{\partial F(\tau_i; \boldsymbol{\theta}_W)}{\partial w_2} = -\frac{d_{i0} R(\tau_i; \boldsymbol{\theta}_W) \ln(d_{i0})}{w_2}, \\ d_{1k} &= \left(\frac{\tau_1}{\beta_1}\right)^{w_2} x_1^k, \\ d_{2k} &= \left(\frac{\tau_1}{\beta_1} + \frac{\tau_2 - \tau_1}{\beta_2}\right)^{w_2 - 1} \left\{\left(\frac{\tau_1}{\alpha_1}\right) x_1^k + \left(\frac{\tau_2 - \tau_1}{\beta_2}\right) x_2^k\right\}. \end{aligned}$$

With this notation, the Fisher information matrix, $I_{obs}(\boldsymbol{\theta})$ in (8.5), is given by

$$I_{obs}(\boldsymbol{\theta}_W) = K \begin{bmatrix} r_{00} + s_{00}\pi & r_{10} + s_{10}\pi & r_{20} + s_{20}\pi \\ r_{10} + s_{10}\pi & r_{11} + s_{11}\pi & r_{21} + s_{21}\pi \\ r_{20} + s_{20}\pi & r_{21} + s_{21}\pi & r_{22} + s_{22}\pi \end{bmatrix}, \tag{8.17}$$

As compared to the exponential case, we observe in this Weibull case that we have three model parameters to be estimated, but with the number of failures to be observed only at two inspection times under a simple SSALT. So with the number of model parameters being more than the number of observational information points, the information matrix becomes singular; consequently, the asymptotic variance–covariance matrix of the MLE of model parameter $\boldsymbol{\theta}_W$ cannot be obtained. The formal proof for the model parameter being nonidentifiable in this case is presented in Appendix C. To handle this situation, therefore, we consider two different options as elaborated in Sections 8.3.3 and 8.3.4.

8.3.3 With a Known Shape Parameter w_2

It is well-known that shape parameter of Weibull distribution corresponds to the failure rate behavior: it being larger than 1 corresponding to an increasing failure rate, less than 1 corresponding to a decreasing failure rate, and equal to 1 corresponding to a constant failure rate (namely, an exponential distribution). Research on Weibull distribution with a known shape parameter has been discussed extensively in reliability literature (see Pascual 2007; Zhang and Meeker 2005; Aslam and Jun 2009, for example). So for this reason, as our first option, we consider Weibull distribution with a known shape parameter w_2, and regard $\mathbf{w}_1 = (w_0, w_1)$ as the model parameter to be estimated. The Fisher information matrix in this case is a 2×2 matrix, from which the asymptotic variance–covariance matrix of the MLE of model parameter is readily obtained to be

$$\begin{aligned} V_{\mathbf{w}_1} &= \frac{1}{K} \begin{bmatrix} r_{00} + s_{00}\pi & r_{10} + s_{10}\pi \\ r_{10} + s_{10}\pi & r_{11} + s_{11}\pi \end{bmatrix}^{-1} \\ &= \frac{1}{D_1} \begin{bmatrix} r_{11} + s_{11}\pi & -(r_{10} + s_{10}\pi) \\ -(r_{10} + s_{10}\pi) & r_{00} + s_{00}\pi \end{bmatrix}, \end{aligned} \tag{8.18}$$

where $D_1 = K\{(r_{00} + s_{00}\pi)(r_{11} + s_{11}\pi) - (r_{10} + s_{10}\pi)^2\}$.

Subsequently, the asymptotic variance of the MLE of reliability at mission time t at normal operating condition x_0 can be obtained by the use of delta method as follows:

$$V_R(\mathbf{w}_1) = P'_{R_1} V_{\mathbf{w}_1} P_{R_1} = \frac{(R(t))^2 w_2^2 d_0^2 (r_{11} + s_{11}\pi)}{K(r_{11}s_{00} + r_{00}s_{11} - 2r_{10}s_{10})\pi(1-\pi)}, \tag{8.19}$$

where $d_0 = (t/\beta_0)^{w_2}$ and $P_{R_1} = [\partial R(t)/\partial w_0, \partial R(t)/\partial w_1]_{2\times 1}$ is a vector of the first-order derivatives of reliability function with respect to model parameters w_0 and w_1.

8.3.4 With a Known Parameter About Stress Level w_1

Another option would be to consider that the parameter about stress level, w_1, is known. For example, the activation energy for humidity in Arrhenius model is known (Vassilious and Mettas, 2001). Alternatively, when the mean lifetimes, $E[T_i] < E[T_j]$, at two stress levels, $0 < x_j < x_i < 1$, could either be suggested by experts or obtained from previous similar studies, this parameter w_1 can once again be known. To be specific, let T_m be the lifetime having Weibull distribution with shape parameter w_2 and scale parameter $\beta_m = \exp(w_0 + w_1 x_m)$, where $m \in \{i, j\}$, corresponding to the two stress levels. Then, by using the expression of mean lifetime given by

$$E[T_m] = \beta_m \Gamma\left(1 + \frac{1}{w_2}\right),$$

we readily find

$$\ln\left(\frac{E[T_i]}{E[T_j]}\right) = \ln\left(\frac{\beta_i}{\beta_j}\right) = w_1(x_i - x_j).$$

Hence, the parameter about stress level w_1 can be expressed as

$$w_1 = \frac{\ln(E[T_i]) - \ln(E[T_j])}{x_i - x_j}.$$

In this case, it would be reasonable to fix w_1 as a known parameter, and then consider the Fisher information matrix of the MLE of model parameter $\mathbf{w}_2 = (w_0, w_2)$. The asymptotic variance–covariance matrix of the MLE of model parameter, in this case, can be shown to be

$$\begin{aligned} V_{\mathbf{w}_2} &= \frac{1}{K}\begin{bmatrix} r_{00} + s_{00}\pi & r_{20} + s_{20}\pi \\ r_{20} + s_{20}\pi & r_{22} + s_{22}\pi \end{bmatrix}^{-1} \\ &= \frac{1}{D_2}\begin{bmatrix} r_{22} + s_{22}\pi & -(r_{20} + s_{20}\pi) \\ -(r_{20} + s_{20}\pi) & r_{00} + s_{00}\pi \end{bmatrix}, \end{aligned} \tag{8.20}$$

where $D_2 = K\{(r_{22} + s_{22}\pi)(r_{00} + s_{00}\pi) - (r_{20} + s_{20}\pi)^2\}$. The asymptotic variance of the MLE of reliability at mission time t can be shown from (8.20), by the use of delta method, to be

$$\begin{aligned} V_R(\mathbf{w}_2) &= P'_{R_2} V_{\mathbf{w}_2} P_{R_2} \\ &= \frac{(R(t)d_0)^2((w_2^2 r_{22} + 2w_2 d_2 r_{20} + d_2^2 r_{00}) + (w_2^2 s_{22} + 2w_2 d_2 s_{20} + d_2^2 s_{00})\pi)}{K(r_{00}s_{22} + r_{22}s_{00} - 2r_{20}s_{20})\pi(1-\pi)}, \end{aligned} \tag{8.21}$$

where $d_0 = (t/\beta_0)^{w_2}$, $d_2 = \ln(t/\beta_0)$, and $P_{R_2} = [\partial R(t)/\partial w_0, \partial R(t)/\partial w_2]_{2\times 1}$ is a vector of the first-order derivatives of reliability function with respect to the model parameters w_0 and w_2.

Moreover, from the asymptotic normality of the MLE $\hat{\theta}$ of model parameter θ, the delta method also provides asymptotic normality for the MLE of a parameter of interest ϕ; specifically, $\hat{\phi} \sim N(\phi, \hat{V}_\phi)$, where $\hat{V}_\phi$ is V_ϕ evaluated at $\hat{\theta}$ and $se(\hat{\phi}) = \sqrt{\hat{V}_\phi}$ is the standard error of the MLE of ϕ.

8.4 Optimal Design of Simple SSALT

We now set to determine the optimal setting of design parameters, such as inspection times and sample allocation, by minimizing some suitable criterion, such as the asymptotic variance of the MLE of a parameter of interest ϕ. For this purpose, upon rewriting the expression of the variance in (8.13), (8.19), and (8.21) as

$$V_\phi = \frac{c_1 + c_2\pi}{K\pi(1-\pi)} = \frac{1}{K}\left(\frac{c_1}{\pi} + \frac{c_1 + c_2}{1-\pi}\right) \tag{8.22}$$

and then minimizing it with respect to π, the proportion of devices to be inspected at time τ_1, we readily find the optimal solution of π to be

$$\pi = \left(1 + \sqrt{\frac{c_1 + c_2}{c_1}}\right)^{-1}. \tag{8.23}$$

Then, the optimal proportions of devices to be inspected at time τ_1, which minimize the asymptotic variance of the MLE of mean life for exponential distribution in (8.13), and the MLE of reliability at mission time t for Weibull distribution in (8.19) and (8.21) can be shown to be

$$\pi_E = \left\{1 + \sqrt{\frac{A_1(X_{11} - X_{10}x_0)^2}{A_2(X_{21} - X_{20}x_0)^2}}\right\}^{-1}, \tag{8.24}$$

$$\pi_1 = \left\{1 + \left(\frac{X_{11}}{X_{21}}\right)\sqrt{\frac{A_1}{A_2}}\right\}^{-1}, \tag{8.25}$$

$$\pi_2 = \left\{1 + \left(\frac{w_2X_{12} + d_2X_{10}}{w_2X_{22} + d_2X_{20}}\right)\sqrt{\frac{A_1}{A_2}}\right\}^{-1}. \tag{8.26}$$

Observe, in this setup, that the percentage π_1 involves X_{21}, which relies on $(x_1, x_2, \tau_1, \tau_2)$ but is independent of mission time t and normal operating condition x_0, whereas π_2 is dependent on mission time t and normal operating condition x_0.

Also, upon substituting the expressions of π_E in (8.24), π_1 in (8.25) and π_2 in (8.26) are substituted into (8.13), (8.19), and (8.21), respectively, we obtain the standard deviations of the MLEs of mean lifetime and reliability at mission time t to be

$$s_\mu = \sqrt{V_\mu} = \frac{1}{\sqrt{K}}\left(\frac{\sum_{i=1}^{2}\sqrt{A_i}(X_{i1} - X_{i0}x_0)^2}{\sqrt{A_1A_2}\lambda_0(X_{10}X_{21} - X_{11}X_{20})}\right) = \frac{C_E}{\sqrt{K}}, \tag{8.27}$$

$$s_{R_1} = \sqrt{V_R(\mathbf{w}_1)} = \frac{1}{\sqrt{K}}\left(\frac{R(t)w_2d_0\left(\sum_{i=1}^{2}\sqrt{A_i}X_{i1}\right)}{\sqrt{A_1A_2}(X_{10}X_{21} - X_{11}X_{20})}\right) = \frac{C_1}{\sqrt{K}}, \tag{8.28}$$

$$s_{R_2} = \sqrt{V_R(\mathbf{w}_2)} = \frac{1}{\sqrt{K}}\left(\frac{R(t)d_0(\sum_{i=1}^{2}\sqrt{A_i}(w_2X_{i2} + d_2X_{i0}))}{\sqrt{A_1A_2}(X_{12}X_{20} - X_{10}X_{22})}\right) = \frac{C_2}{\sqrt{K}}. \tag{8.29}$$

Observe that the standard deviations of the MLEs are inversely proportional to the square root of the sample size K, implying that a smaller standard deviation will be attained when sample size increases. Also, as C_E, C_1, and C_2 are nonlinear functions of τ_1 and τ_2, some nonlinear optimizing tools, such as **fminsearch** in Matlab, or **optim** in R, can be used for minimizing (8.27), (8.28), and (8.29). We may then adopt the following procedure for determining the inspection time for each of the two stages of the SSALT, subject to a specification of the maximum experimental time τ_{ter} and stress levels x_1 and x_2:

Step 1: Define $\tau_i = \tau_{\text{ter}}\{1 - \exp(-\sum_{k=1}^{i}\exp(\upsilon_k))\}$, for $i = 1, 2$;
Step 2: Find (υ_1, υ_2) that minimizes C_E in (8.27) or C_1 in (8.28) or C_2 in (8.29) by using an optimization tool;
Step 3: Compute (τ_1, τ_2) with the values of (υ_1, υ_2) determined in Step 2.

Let $C = \{C_E, C_1, C_2\}$. As $0 < \exp(-(\exp(\upsilon_1) + \exp(\upsilon_2))) < \exp(-(\exp(\upsilon_1))) < 1$ for $-\infty < \upsilon_1 < \infty$ and $-\infty < \upsilon_2 < \infty$, Step 1 would guarantee that $0 < \tau_1 < \tau_2 < \tau_{\text{ter}}$. In addition, given the limit on standard deviation of the MLE of a parameter of interest SD, the required sample size becomes

$$K \geq \left(\frac{C}{SD}\right)^2.$$

Then, π_E can be determined from (8.24) with $(x_0, A_1, A_2, X_{10}, X_{11}, X_{20}, X_{21})$, π_1 can be determined from (8.25) with $(A_1, A_2, X_{11}, X_{21})$, and π_2 can be determined from (8.26) with $(w_2, t, x_0, A_1, A_2, X_{12}, X_{10}, X_{22}, X_{20})$, from which we can obtain $K_1 = K\pi$ and $K_2 = K - K_1$. Note that C_1 in (8.28) is a nonlinear function of x_1, x_2, τ_1, τ_2; but υ_1 and υ_2 in Step 1 will be the same as long as $(x_1, x_2, \tau_{\text{ter}})$ remain the same, regardless of the limit on standard deviation, the mission time, and the normal operating condition; thus, the inspection times τ_1 and τ_2 will also remain the same. However, the limit on standard deviation, the mission time, and the normal operating condition would affect the required sample size.

8.5 Case Studies with R Codes

8.5.1 SSALT for Exponential Distribution

Fan et al. (2009) considered exponential distribution for analyzing data on electro-explosive devices collected from CSALTs. We now present the optimal design of simple SSALTs for electro-explosive devices. Suppose a simple SSALT is to be conducted for estimating the mean lifetime of devices under the normal operating condition of 298 K. Suppose the parameter values (e_0^*, e_1^*) are specified to be $(11.7, 4750)$ with $\mu = 69.38$, for the purpose of devising an optimal SSALT plan. It is worth noting that the predictor $1/Temp$ is usually considered in reliability studies, so x_0, x_1, and x_2 were set to be the negative of the reciprocal of temperature (K) such that $x_0 < x_1 < x_2$. The R codes for the above algorithm, for some specifications, are presented in Table 8.2.

Table 8.3 presents the optimal designs of simple SSALTs for the electro-explosive devices under different settings. A simulation study, for each of the determined optimal designs, is also carried out (based on 1000 experiments) to find the corresponding empirical mean, $\bar{x}_{\hat{\mu}}$, standard deviation, $s_{\hat{\mu}}$, and standard errors, $se(\hat{\mu})$, of the estimate of mean lifetime, $\hat{\mu}$. These values are presented in the last three columns of Table 8.3. We observe that the empirical standard deviations of the estimate of mean lifetime and the empirical standard errors are close to the standard deviations pre-fixed for the plans. We also observe that the empirical means of the estimate of mean lifetime are slightly larger than the true value. We observe,

Table 8.2 R codes for optimal SSALTs for the exponential distribution.

```
source("SSOTP.R")
e0<-11.7; e1<-4750
x1<- -1/308 #stress level at 1st stage
x2<- -1/323 #stress level at 2nd stage
x0<- -1/298 #normal operating condition
TT<-15 #termination time
SD<-10 #limit of standard deviation
SSOTP<-SSALT.exp(e0,e1,x1,x2,x0,TT,SD)
```

Table 8.3 Optimal designs of SSALTs for electro-explosive devices with parameter values $(e_0^*, e_1^*) = (11.7, 4750)$ and mean lifetime $\mu = 69.38$.

Setting				SSALT plan						
SD	τ_{ter}	$-1/x_1$	$-1/x_2$	τ_1	τ_2	K_1	K_2	$\bar{x}_{\hat{\mu}}$	$s_{\hat{\mu}}$	$se(\hat{\mu})$
10	15	308	323	6.81	15	1 681	507	69.79	10.07	10.09
10	30	308	323	13.06	30	948	313	70.50	10.50	10.19
10	60	308	323	24.04	60	612	257	70.30	9.90	10.15
10	15	308	328	7.65	15	1 138	301	70.00	10.20	10.13
10	30	308	328	14.75	30	649	192	69.79	10.04	10.08
10	60	308	328	27.67	60	431	172	70.60	10.18	10.22
15	15	308	323	6.81	15	748	225	71.52	15.01	15.54
15	30	308	323	13.06	30	422	139	71.40	15.96	15.47
15	60	308	323	24.04	60	272	114	71.09	15.02	15.46
15	15	308	328	7.65	15	506	134	71.17	15.29	15.50
15	30	308	328	14.75	30	289	85	71.49	16.59	15.55
15	60	308	328	27.67	60	192	76	72.11	16.50	17.79
10	15	318	323	4.28	15	20 794	9 999	69.89	9.52	10.07
10	30	318	323	7.89	30	12 111	6 588	69.58	9.99	10.03
10	60	318	323	13.15	59	8 485	6 285	69.71	9.86	10.05
10	15	318	328	5.05	15	5 884	2 554	70.43	10.10	10.15
10	30	318	328	9.35	30	3 506	1 776	70.30	9.96	10.13
10	60	318	328	14.35	51	2 689	1 808	70.05	9.88	10.10
15	15	318	323	4.28	15	9 242	4 444	71.33	15.32	15.41
15	30	318	323	7.89	30	5 383	2 928	71.53	14.87	15.46
15	60	318	323	13.15	59	3 771	2 794	70.51	14.78	15.25
15	15	318	328	5.05	15	2 615	1 135	71.51	15.60	15.46
15	30	318	328	9.35	30	1 559	789	70.97	15.43	15.34
15	60	318	328	14.35	51	1 195	804	71.36	16.39	15.45

Here, the normal operating condition is 298 K.

Table 8.4 Sensitivity analysis of the choice of values (e_0^*, e_1^*) with small ($\pm 0.5\%$)/moderate ($\pm 1\%$) errors to the design of simple SSALTs, along with the values of corresponding empirical mean $\bar{x}_{\hat{\mu}}$, standard deviation $s_{\hat{\mu}}$ and standard error $se(\hat{\mu})$ of the MLE of mean lifetime $\hat{\mu}$.

ϵ_1	ϵ_2	e_0^*	e_1^*	μ	$-1/x_1$	$-1/x_2$	τ_1	τ_2	K_1	K_2	$\bar{x}_{\hat{\mu}}$	$s_{\hat{\mu}}$	$se(\hat{\mu})$
0	0	11.70	4750	69.38	308	328	7.65	15	506	134	71.96	16.45	15.67
−0.005	−0.005	11.64	4726	67.92	308	328	7.64	15	478	127	72.74	16.07	16.29
−0.005	0	11.64	4750	73.56	308	328	7.67	15	598	158	70.58	14.00	14.13
−0.005	0.005	11.64	4774	79.66	308	328	7.70	15	750	196	70.20	12.86	12.52
0	−0.005	11.70	4726	64.07	308	328	7.62	15	404	108	72.10	18.53	17.62
0	0.005	11.70	4774	75.14	308	328	7.68	15	633	167	70.78	13.70	13.78
0.005	−0.005	11.76	4726	60.42	308	328	7.60	15	342	92	72.45	19.29	19.23
0.005	0	11.76	4750	65.44	308	328	7.64	15	427	114	72.16	18.00	17.15
0.005	0.005	11.76	4774	70.87	308	328	7.67	15	535	142	71.02	14.52	15.01
−0.01	−0.01	11.58	4703	66.50	308	328	7.62	15	452	120	71.99	17.07	16.63
−0.01	0	11.58	4750	77.99	308	328	7.68	15	708	186	70.81	13.73	13.04
−0.01	0.01	11.58	4798	91.47	308	328	7.74	15	1114	288	69.89	9.96	10.24
0	−0.01	11.70	4703	59.16	308	328	7.59	15	323	87	72.46	19.91	19.85
0	0	11.70	4750	69.38	308	328	7.65	15	506	134	71.69	15.66	15.60
0	0.01	11.70	4798	81.37	308	328	7.71	15	794	207	70.08	12.31	12.18
0.01	−0.01	11.82	4703	52.63	308	328	7.54	15	232	63	73.97	25.45	24.04
0.01	0	11.82	4750	61.72	308	328	7.62	15	362	97	72.80	19.47	18.78
0.01	0.01	11.82	4798	72.38	308	328	7.68	15	567	149	69.59	14.57	14.33

Here, the normal operating condition is 298 K.

in addition, that when the standard deviation is fixed, increasing the stress level at the second stage x_2 results in reducing the required sample sizes significantly. The results in Table 8.3 also show that $\tau_2 < \tau_{\text{Ter}}$ in some settings, implying that when the stress level at the second stage x_2 is set to be sufficiently high, increasing the number of failures by prolonging the experimental time τ_{ter} does not provide more information for the estimation of mean lifetime. We also observe that, when the normal operating condition is 298 K, if the stress level at the first stage is increased from 308 to 318 K (away from the normal operating condition), then the corresponding optimal plan would require a much larger sample size for the same specification of termination time and the limit on standard deviation of the MLE of mean lifetime of devices.

As in Chapter 7, it is important to examine the effect of misspecification of the parameter values of (e_0^*, e_1^*) on the optimal SSALT plans in order to evaluate their robustness properties. Because the specified values of (e_0^*, e_1^*) are likely to depart from the true model parameters (e_0, e_1), we assume now that $(e_0^*, e_1^*) = (e_0(1 + \epsilon_1), e_1(1 + \epsilon_2))$, where $\epsilon_i \in \{-0.01, -0.005, 0.00, 0.005, 0.01\}$, thus allowing for underspecification as well as overspecification of the true values of the model parameters. We further assume that $(x_1, x_2, \tau_{\text{ter}}, s_\mu)$ are set to be $(-1/308, -1/328, 30, 15)$. Table 8.4 presents the results of the sensitivity analysis on the design of simple SSALTs, along with the corresponding simulated mean $\bar{x}_{\hat{\mu}}$ and standard deviation $s_{\hat{\mu}}$ of the MLE of mean lifetime under the normal operating condition of 298 K. We observe that, within small ($\pm$0.5%) and moderate ($\pm$1%) departures of (e_0, e_1), the optimal design of simple SSALTs are robust only when the estimated mean is close to the true mean. In general, when the stress levels and the termination time are constant, the inspection times at the two stress levels, τ_1 and τ_2, remain almost the same among the designs. But the required sample size highly depends on the estimated mean lifetime. For example, when ϵ_1 increases by 1%, the estimated mean lifetime is much larger than the actual mean lifetime, resulting in a larger required sample size.

8.5.2 SSALT for Weibull Distribution

In this section, we present an application of the optimal design for G-MRFs. Zheng et al. (2018) considered Weibull distribution for analyzing data on G-MRFs samples collected from a SSALT with four stress levels. In this test, the exact failure time of each G-MRF sample is not observed and only whether its viscosity or shear stress has decreased by more than 10% at an inspection time is observed. This results in one-shot device testing data. In their study, they observed that the shape parameters were similar at those four levels. We then assume, based on their results, that $w_0 = -24.654, w_1 = -10\,103.698, w_2 = 2.3$ for determining optimal SSALTs for G-MRFs samples. Suppose the lowest and the highest temperatures that could be used are 333 and 351 K, respectively, and that we wish to estimate the reliability of G-MRFs at mission time $t = 14\,588$ hours at a temperature of 293 K, and that we specify the limit on standard deviation of the MLE of reliability to be 0.1. The R codes for the algorithms described in Section 8.4, for some specifications, are presented in Table 8.5.

Table 8.5 R codes for optimal SSALTs for the Weibull distribution.

```
source("SSOTP.R")
w0<- -24.654; w1<- -10 103.698; w2<-2.3
x1<- -1/333 #stress level at 1st stage
x2<- -1/351 #stress level at 2nd stage
x0<- -1/293 #normal operating condition
TT<-200 #termination time
SD<-0.1 #limit of standard deviation
SSOTP1<-SSALT.Wei(w0,w1,w2,x1,x2,x0,TT,SD,"s1") # with fixed w1
SSOTP2<-SSALT.Wei(w0,w1,w2,x1,x2,x0,TT,SD,"s2") # with fixed w2
```

Table 8.6 Optimal designs for G-MRFs (Zheng et al., 2018) to estimate the reliability at mission time $t = 14\ 588$ hours at a temperature of 293 K.

	w_1 is assumed known[a)]		
Stage (i)	**Temperature (K)**	**τ_i (h)**	**K_i**
1	333	125	0
2	351	147	25
	w_2 is assumed known[b)]		
Stage (i)	**Temperature (K)**	**τ_i (h)**	**K_i**
1	333	141	2236
2	351	200	945

a) SSOTP1.
b) SSOTP2.

For the case when the parameter about stress level is assumed known, i.e. $w_1 = 10\ 103.698$, the optimal design of a SSALT is presented in Table 8.6. It suggests that a sample of 25 G-MRFs is sufficient for this reliability estimation and that the temperature is to increase from 333 to 351 K at 125 hours and that all the G-MRFs are to be inspected at 147 hours. Observe that this optimal design takes less experimental time than 864 hours need for the original experiment of

Zheng et al. (2018). In addition, for the case when the shape parameter is assumed known, i.e. $w_2 = 2.3$, the required sample size is much larger.

Furthermore, we present the results of some simulation studies to examine the properties of optimal designs described for one-shot devices with Weibull lifetime distribution with various failure rates. We consider values for the scale parameter to be $(w_0, w_1) = (5.5, -0.05)$, and a range of shape parameter to be $w_2 \in \{0.8, 1, 2\}$ for representing decreasing ($w_2 < 1$), constant ($w_2 = 1$) and increasing ($w_2 > 1$) failure rates. In addition, the reliability estimation under normal operating condition $x_0 = 25$ at mission time $t = 60$ is considered for the determination of optimal designs of simple SSALTs for one-shot devices.

The determined optimal designs, for the limit on standard deviation $SD \in \{0.01, 0.05\}$ and different maximum experimental times τ_{ter}, are presented in Tables 8.7 and 8.8 for the case of fixed w_2 and in Tables 8.9 and 8.10 for the case of fixed w_1. Based on 1000 Monte Carlo simulations, the empirical mean, $\bar{x}_{\hat{R}}$, and standard deviation, $s_{\hat{R}}$, of the estimate of reliability at the mission time are also presented in the last two columns of these tables.

We see from the obtained results that the means of the estimate of reliability under the determined optimal designs are close to the true values. We also observe that the determined optimal designs, for the case when w_2 is fixed, can maintain standard deviation at specified levels only when the number of tested devices is large ($K > 500$). Moreover, if the specified limit on standard deviation of the MLE of reliability gets increased, then required sample size decreases, but the inspection times remain unchanged. Finally, we observe in some cases that $\tau_2 \neq \tau_{\text{ter}}$, implying that an unnecessarily long period of experimental time does not result in reducing sample size.

For optimal designs, for the case when w_1 is fixed, the simulation studies reveal that the determined optimal designs do maintain the standard deviations at specified levels in all considered cases. Also, in this setup, either all the devices get inspected at the first inspection time when the experimental time is allowed to be sufficiently large, or all the devices get inspected at the second inspection time when the experimental time is made shorter. It needs to be mentioned, in this case, that stress levels affect the inspection times, but do not have any effect on sample size. A formal proof of this property is presented in Appendix D. Finally, in this case when the parameter about stress level w_1 is known, the required sample size is seen to be smaller and the empirical standard deviation of the estimate of reliability to be robust, compared to case when the shape parameter is assumed known.

Furthermore, a sensitivity analysis is carried out to examine the effect of misspecification of parameter values on the optimal designs. As done before, we assume that the parameter values (w_0^*, w_1^*, w_2^*) differ from the true values of the model parameters (w_0, w_1, w_2) with small (± 0.01)/moderate (± 0.05) errors

Table 8.7 Optimal designs with fixed w_2 and limit on standard deviation of 0.05 for one-shot devices with various failure rates, along with the values of corresponding empirical mean and standard deviation of the estimate of reliability.

	Setting				SSALT plan					
w_2	x_1	x_2	τ_{ter}	SD	K_1	K_2	τ_1	τ_2	$\bar{x}_{\hat{R}}$	$s_{\hat{R}}$
2[a)]	35	55	50	0.05	588	278	32.54	49.04	0.482	0.050
2[a)]	35	55	40	0.05	725	239	26.06	40.00	0.482	0.050
2[a)]	35	55	30	0.05	1110	281	19.64	30.00	0.480	0.050
2[a)]	30	55	70	0.05	240	79	48.92	64.42	0.497	0.068
2[a)]	30	55	60	0.05	252	71	45.16	60.00	0.489	0.060
2[a)]	30	55	50	0.05	307	65	37.33	50.00	0.483	0.050
1[b)]	35	55	80	0.05	407	175	33.48	70.36	0.427	0.050
1[b)]	35	55	70	0.05	407	175	33.31	70.00	0.428	0.050
1[b)]	35	55	60	0.05	429	163	28.87	60.00	0.426	0.050
1[b)]	30	55	100	0.05	189	51	55.15	92.78	0.447	0.071
1[b)]	30	55	90	0.05	190	50	53.58	90.00	0.442	0.069
1[b)]	30	55	80	0.05	197	47	48.01	80.00	0.435	0.063
0.8[c)]	35	55	90	0.05	370	155	35.05	85.76	0.416	0.051
0.8[c)]	35	55	80	0.05	371	151	33.01	80.00	0.417	0.050
0.8[c)]	35	55	70	0.05	389	145	29.41	70.00	0.416	0.051
0.8[c)]	30	55	120	0.05	178	45	59.70	113.18	0.437	0.072
0.8[c)]	30	55	110	0.05	179	44	58.23	110.00	0.435	0.072
0.8[c)]	30	55	100	0.05	183	42	53.63	100.00	0.430	0.069

a) R(60)=0.481.
b) R(60)=0.425.
c) R(60)=0.414.

Table 8.8 Optimal designs with fixed w_2 and limit on standard deviation of 0.01 for one-shot devices with various failure rates, along with the values of corresponding empirical mean and standard deviation of the estimate of reliability.

	Setting				SSALT plan					
w_2	x_1	x_2	τ_{ter}	SD	K_1	K_2	τ_1	τ_2	$\bar{x}_{\hat{R}}$	$s_{\hat{R}}$
2[a)]	35	55	50	0.01	14 699	6 940	32.54	49.04	0.481	0.010
2[a)]	35	55	40	0.01	18 116	5 982	26.06	40.00	0.481	0.010
2[a)]	35	55	30	0.01	27 748	7 024	19.64	30.00	0.481	0.010
2[a)]	30	55	70	0.01	5 998	1 965	48.92	64.42	0.481	0.010
2[a)]	30	55	60	0.01	6 302	1 779	45.16	60.00	0.481	0.010
2[a)]	30	55	50	0.01	7 671	1 617	37.33	50.00	0.481	0.010
1[b)]	35	55	80	0.01	10 163	4 385	33.48	70.36	0.425	0.010
1[b)]	35	55	70	0.01	10 177	4 371	33.31	70.00	0.425	0.010
1[b)]	35	55	60	0.01	10 735	4 074	28.87	60.00	0.425	0.010
1[b)]	30	55	100	0.01	4 733	1 271	55.15	92.78	0.425	0.010
1[b)]	30	55	90	0.01	4 764	1 244	53.58	90.00	0.425	0.010
1[b)]	30	55	80	0.01	4 929	1 166	48.01	80.00	0.425	0.010
0.8[c)]	35	55	90	0.01	9 249	3 874	35.05	85.76	0.414	0.010
0.8[c)]	35	55	80	0.01	9 388	3 764	33.01	80.00	0.414	0.010
0.8[c)]	35	55	70	0.01	9 733	3 624	29.41	70.00	0.413	0.010
0.8[c)]	30	55	120	0.01	4 457	1 124	59.70	113.18	0.414	0.010
0.8[c)]	30	55	110	0.01	4 476	1 107	58.23	110.00	0.414	0.010
0.8[c)]	30	55	100	0.01	4 557	1 061	53.63	100.00	0.414	0.010

a) R(60)=0.481.
b) R(60)=0.425.
c) R(60)=0.414.

Table 8.9 Optimal designs with fixed w_1 and limit on standard deviation of 0.05 for one-shot devices with various failure rates, along with the values of corresponding empirical mean and standard deviation of the estimate of reliability.

	Setting				SSALT plan					
w_2	x_1	x_2	τ_{ter}	SD	K_1	K_2	τ_1	τ_2	$\bar{x}_{\hat{R}}$	$s_{\hat{R}}$
2[a)]	35	55	50	0.05	100	0	36.39	39.43	0.480	0.050
2[a)]	35	55	40	0.05	0	100	25.31	29.39	0.481	0.049
2[a)]	35	55	30	0.05	0	100	11.49	20.65	0.481	0.050
2[a)]	30	55	60	0.05	100	0	46.73	48.75	0.481	0.050
2[a)]	30	55	50	0.05	0	100	36.42	39.38	0.481	0.050
2[a)]	30	55	40	0.05	0	100	19.94	27.61	0.481	0.051
1[b)]	35	55	50	0.05	98	0	36.39	39.49	0.426	0.050
1[b)]	35	55	40	0.05	0	98	25.29	29.37	0.425	0.050
1[b)]	35	55	30	0.05	0	98	12.04	21.00	0.425	0.050
1[b)]	30	55	60	0.05	98	0	46.73	48.74	0.425	0.050
1[b)]	30	55	50	0.05	0	98	36.40	39.36	0.425	0.050
1[b)]	30	55	40	0.05	0	98	19.50	27.30	0.425	0.050
0.8[c)]	35	55	50	0.05	97	0	36.39	39.63	0.414	0.050
0.8[c)]	35	55	40	0.05	0	97	25.29	29.37	0.413	0.050
0.8[c)]	35	55	30	0.05	0	97	12.13	21.06	0.413	0.050
0.8[c)]	30	55	60	0.05	97	0	46.73	48.75	0.413	0.050
0.8[c)]	30	55	40	0.05	0	97	36.40	39.36	0.413	0.050
0.8[c)]	30	55	30	0.05	0	97	19.50	27.30	0.412	0.050

a) R(60)=0.481.
b) R(60)=0.425.
c) R(60)=0.414.

Table 8.10 Optimal designs with fixed w_1 and limit on standard deviation of 0.01 for one-shot devices with various failure rates, along with the values of corresponding empirical mean and standard deviation of the estimate of reliability.

	Setting				SSALT plan					
w_2	x_1	x_2	τ_{ter}	SD	K_1	K_2	τ_1	τ_2	$\bar{x}_{\hat{R}}$	$s_{\hat{R}}$
2[a)]	35	55	50	0.01	2496	0	36.39	39.43	0.481	0.010
2[a)]	35	55	40	0.01	0	2496	25.31	29.39	0.481	0.010
2[a)]	35	55	30	0.01	0	2496	11.49	20.65	0.481	0.010
2[a)]	30	55	60	0.01	2496	0	46.73	48.75	0.480	0.010
2[a)]	30	55	50	0.01	0	2496	36.42	39.38	0.481	0.010
2[a)]	30	55	40	0.01	0	2496	19.94	27.61	0.481	0.010
1[b)]	35	55	50	0.01	2444	0	36.39	39.49	0.425	0.010
1[b)]	35	55	40	0.01	0	2444	25.29	29.37	0.425	0.010
1[b)]	35	55	30	0.01	0	2444	12.04	21.00	0.425	0.010
1[b)]	30	55	60	0.01	2444	0	46.73	48.74	0.425	0.010
1[b)]	30	55	50	0.01	0	2444	36.40	39.36	0.425	0.010
1[b)]	30	55	40	0.01	0	2444	19.50	27.30	0.425	0.010
0.8[c)]	35	55	50	0.01	2425	0	36.39	39.63	0.414	0.010
0.8[c)]	35	55	40	0.01	0	2425	25.29	29.37	0.414	0.010
0.8[c)]	35	55	30	0.01	0	2425	12.13	21.06	0.414	0.010
0.8[c)]	30	55	60	0.01	2425	0	46.73	48.75	0.414	0.010
0.8[c)]	30	55	40	0.01	0	2425	36.40	39.36	0.414	0.010
0.8[c)]	30	55	30	0.01	0	2425	19.50	27.30	0.414	0.010

a) R(60)=0.481.
b) R(60)=0.425.
c) R(60)=0.414.

Table 8.11 Sensitivity analysis of the choice of parameter values (w_0^*, w_1^*, w_2^*) with small/moderate errors ϵ_1 on optimal designs with fixed w_2, along with the values of corresponding misspecified reliability $R^*(t)$.

ϵ_1	ϵ_2	ϵ_3	w_0^*	w_1^*	w_2^*	τ_1	τ_2	K_1	K_2	$R^*(t)$
0	**0**	**0**	**5.5**	**-0.05**	**2**	**19.64**	**30**	**1110**	**281**	**0.481**
−0.01	−0.01	−0.01	5.445	−0.0495	1.98	19.52	30	1082	281	0.450
−0.01	−0.01	0.01	5.445	−0.0495	2.02	19.64	30	1103	286	0.451
−0.01	0.01	−0.01	5.445	−0.0505	1.98	19.58	30	1032	275	0.432
−0.01	0.01	0.01	5.445	−0.0505	2.02	19.70	30	1052	279	0.433
0.01	−0.01	−0.01	5.555	−0.0495	1.98	19.59	30	1145	280	0.528
0.01	−0.01	0.01	5.555	−0.0495	2.02	19.70	30	1167	284	0.529
0.01	0.01	−0.01	5.555	−0.0505	1.98	19.63	30	1106	275	0.509
0.01	0.01	0.01	5.555	−0.0505	2.02	19.75	30	1127	279	0.511
−0.05	−0.05	−0.05	5.225	−0.0475	1.9	19.01	30	880	257	0.328
−0.05	−0.05	0.05	5.225	−0.0475	2.1	19.61	30	953	276	0.324
−0.05	0.05	−0.05	5.225	−0.0525	1.9	19.56	30	603	224	0.243
−0.05	0.05	0.05	5.225	−0.0525	2.1	20.25	30	619	239	0.231
0.05	−0.05	−0.05	5.775	−0.0475	1.9	19.37	30	1108	251	0.676
0.05	−0.05	0.05	5.775	−0.0475	2.1	19.92	30	1182	260	0.701
0.05	0.05	−0.05	5.775	−0.0525	1.9	19.64	30	1017	238	0.608
0.05	0.05	0.05	5.775	−0.0525	2.1	20.19	30	1093	247	0.630

in the form $(w_0^*, w_1^*, w_2^*) = (w_0(1+\epsilon_1), w_1(1+\epsilon_2), w_2(1+\epsilon_3))$. We further set $(-1/x_0, -1/x_1, -1/x_2, \tau_{ter}, SD) = (25, 35, 55, 30, 0.05)$. Tables 8.11 and 8.12 then present the determined designs for the two cases, namely, when w_2 is assumed known and w_1 is assumed known, respectively, along with the values of the corresponding misspecified reliability, $R^*(t)$.

For the determined designs with fixed w_2, we observe that, with small error ($\pm 1\%$), the determined designs are quite robust in terms of sample sizes and inspection times. But, with moderate error ($\pm 5\%$), sample sizes vary over the determined designs, although the inspection times remain stable. The sample size seems to depend highly on w_0^* that has an impact on the estimated reliability. For the designs with fixed w_1, we observe that the determined designs are quite robust for both small ($\pm 1\%$) and moderate ($\pm 5\%$) errors. Thus, we observe that the optimal designs with fixed w_1 are more robust and require less samples than the corresponding optimal designs

Table 8.12 Sensitivity analysis of the choice of parameter values (w_0^*, w_1^*, w_2^*) with small/moderate errors ϵ_1 on optimal designs with fixed w_1, along with the values of corresponding misspecified reliability $R^*(t)$.

ϵ_1	ϵ_2	ϵ_3	w_0^*	w_1^*	w_2^*	τ_1	τ_2	K_1	K_2	$R^*(t)$
0	**0**	**0**	**5.5**	**−0.05**	**2**	**11.49**	**20.65**	**0**	**100**	**0.481**
−0.01	−0.01	−0.01	5.445	−0.0495	1.98	11.49	20.81	0	99	0.450
−0.01	−0.01	0.01	5.445	−0.0495	2.02	11.49	20.81	0	99	0.451
−0.01	0.01	−0.01	5.445	−0.0505	1.98	11.26	20.35	0	98	0.432
−0.01	0.01	0.01	5.445	−0.0505	2.02	11.26	20.35	0	98	0.433
0.01	−0.01	−0.01	5.555	−0.0495	1.98	11.49	20.81	0	100	0.526
0.01	−0.01	0.01	5.555	−0.0495	2.02	11.49	20.81	0	100	0.529
0.01	0.01	−0.01	5.555	−0.0505	1.98	11.27	20.35	0	100	0.509
0.01	0.01	0.01	5.555	−0.0505	2.02	11.26	20.35	0	100	0.512
−0.05	−0.05	−0.05	5.225	−0.0475	1.9	10.32	20.76	0	88	0.328
−0.05	−0.05	0.05	5.225	−0.0475	2.1	10.32	20.76	0	88	0.324
−0.05	0.05	−0.05	5.225	−0.0525	1.9	12.49	20.53	0	74	0.243
−0.05	0.05	0.05	5.225	−0.0525	2.1	12.69	20.67	0	71	0.231
0.05	−0.05	−0.05	5.775	−0.0475	1.9	10.32	20.76	0	88	0.676
0.05	−0.05	0.05	5.775	−0.0475	2.1	10.32	20.76	0	84	0.701
0.05	0.05	−0.05	5.775	−0.0525	1.9	12.55	20.58	0	95	0.608
0.05	0.05	0.05	5.775	−0.0525	2.1	12.45	20.51	0	93	0.630

with fixed w_2; however, as pointed out earlier, the former ones may carry out inspection of all K units under test only at one stress level, either the lower or upper level, depending on how small or large the maximum test duration is.

9

Competing-Risks Models

9.1 Brief Overview

In this chapter, a competing risks model is considered for one-shot device testing data with multiple failure modes collected from constant-stress accelerated life-tests (CSALTs). The likelihood approach, via expectation-maximization (EM) algorithm and the Bayesian approach with various priors, is discussed for the estimation of model parameters. Due to the consideration of competing risks, the joint posterior distribution becomes quite complicated. Also, two different scenarios are considered for the observed data: first in the case when the observed data contain no masking, and then the case when masking is present in the observed data. The discussions provided here are primarily from the works carried out by Balakrishnan et al. (2015, 2016a, b).

9.2 One-Shot Device Testing Data with Competing Risks

One-shot devices often have multiple components that can cause the failure of the device. For example, a fire extinguisher contains a cylinder, a valve, and chemicals inside; an automobile air bag contains a crash sensor, an inflator, and an air bag; and for any packed food (which is also a kind of one-shot device), there are different causes for food expiry such as the growth of microorganism in the package, the moisture level,and the food deterioration due to oxidation. The failure of any component of a device will result in the failure of the device. However, by the very nature of one-shot devices, each device can only be used once at a specified time and it gets destroyed right after its use. For a failed device, we will normally check for the cause responsible for the failure. Thus, the information collected from life-tests on one-shot devices in this case will include the status of a device

Accelerated Life Testing of One-shot Devices: Data Collection and Analysis, First Edition.
Narayanaswamy Balakrishnan, Man Ho Ling, and Hon Yiu So.

Companion Website: www.wiley.com/go/Balakrishnan/Accelerated_Life_Testing

inspected at a specific time as well as the cause of the failure had the device failed. CSALTs with I test groups for such one-shot devices with G failure modes is set up as follows: For $i = 1, 2, \dots, I$,

1. there are K_i devices placed on test at temperature (as stress level) x_i and inspected at time τ_i;
2. the number of devices failed with the gth failure mode is observed and denoted by $n_{i,g}$, for $g = 1, 2, \dots, G$;
3. the number of surviving devices is observed and denoted by $n_{i,0} = K_i - \sum_{g=1}^{G} n_{i,g}$.

For convenience of discussion, we shall confine our attention here to the case of two failure modes corresponding to the failure of each device. As an example, let us consider the testing of electro-explosive devices; see, for example, Thomas and Betts (1967) and Balakrishnan et al. (2015). Let us now assume that there are only two failure modes for the failure of detonation of the explosive device, say, burnout of resistance wires as failure mode 1 and leakage of organic fuel as failure mode 2. The extension to the case of multiple failure modes can be done in a natural way (with a lot more complicated notation, of course). Let us denote the random time for the gth failure mode by $T_{i,g,k}$, for $g = 1, 2, i = 1, 2, \dots, I$, and $k = 1, 2, \dots, K_i$.

We define $\Delta_{i,k}$ to be the indicator for the kth device under temperature x_i and inspection time τ_i. When the device is successfully detonated, we will set $\Delta_{i,k} = 0$. However, if the device fails to explode, we will identify (by a careful follow-up examination) the specific failure mode responsible for its failure. If the gth failure mode is identified for the failure, we will denote this event by $\Delta_{i,k} = g$, for $g = 1, 2$. Mathematically, the indicator $\Delta_{i,k}$ is then defined as

$$\Delta_{i,k} = \begin{cases} 0 & \text{if } \min(T_{i,1,k}, T_{i,2,k}) > \tau_i, \\ 1 & \text{if } T_{i,1,k} < \min(T_{i,2,k}, \tau_i), \\ 2 & \text{if } T_{i,2,k} < \min(T_{i,1,k}, \tau_i), \end{cases} \tag{9.1}$$

and then $\delta_{i,k}$ will correspond to the realization of $\Delta_{i,k}$. It is important to mention that we are assuming here that we will be able to know exactly which of the two failure modes was responsible for the failure of the device after a careful follow-up examination of the failed device. This form of data is commonly referred to as *autopsy data* in the reliability literature; see, for example the discussions by Meilijson (1981), Nowik (1990), Antoine et al. (1993), among others in this regard.

Now, the data collected at temperatures $\{x_i, i = 1, 2, \dots, I\}$ and the inspection times $\{\tau_i, i = 1, 2, \dots, I\}$ are the numbers of devices with the indicator values $\delta_{i,k} = 0, \delta_{i,k} = 1$ and $\delta_{i,k} = 2$, denoted by $n_{i,0}, n_{i,1}$ and $n_{i,2}$, respectively. The data thus observed from such an experiment can be summarized in the form as in Table 9.1.

Table 9.1 One-shot device testing data collected from CSALTs with I test groups and two failure modes.

Test group	Inspection time	Temperature	Number of devices with $\delta = 0$	$\delta = 1$	$\delta = 2$
1	τ_1	x_1	$n_{1,0}$	$n_{1,1}$	$n_{1,2}$
2	τ_2	x_2	$n_{2,0}$	$n_{2,1}$	$n_{2,2}$
$\vdots$	$\vdots$	$\vdots$	$\vdots$	$\vdots$	$\vdots$
I	τ_I	x_I	$n_{I,0}$	$n_{I,1}$	$n_{I,2}$

9.3 Likelihood Estimation for Exponential Distribution

Here, we assume that the lifetime distribution of $T_{i,g,k}$ is exponential with rate parameter $\lambda_{i,g}$, and then present the EM algorithm for obtaining the maximum likelihood estimates (MLEs) of the model parameters for the competing risks model based on one-shot device testing data with two independent failure modes. This work was originally developed by Balakrishnan et al. (2015).

The corresponding pdf and cdf are given by

$$f_{E,i,g}(t) = \lambda_{i,g} \exp(-\lambda_{i,g}t), \tag{9.2}$$

$$F_{E,i,g}(t) = 1 - \exp(-\lambda_{i,g}t), \quad t > 0, \lambda_{i,g} > 0, \tag{9.3}$$

for $g = 1, 2, i = 1, 2, \dots, I$, where $\lambda_{i,g}$ is the failure rate of the gth component in the device under test at temperature x_i. Of course, $t_{i,g,k}$ will be used to denote the realization of the random variable $T_{i,g,k}$. The relationship between $\lambda_{i,g}$ and x_i is assumed to be a log-linear function of the form

$$\lambda_{i,g} = \exp(e_{g0} + e_{g1}x_i), \tag{9.4}$$

for $g = 1, 2$. Furthermore, under normal operating condition x_0, the mean time for the gth failure mode is given by

$$\mu_{E,g} = \frac{1}{\lambda_{0,g}}, \quad g = 1, 2,$$

while the mean lifetime of device is given by

$$\mu_{ECR} = \frac{1}{\lambda_{0,1} + \lambda_{0,2}},$$

where $\lambda_{0,g} = \exp(e_{g0} + e_{g1}x_0)$, according to (9.4). In this setting, the new model parameter is

$$\theta_{ECR} = (e_{10}, e_{11}, e_{20}, e_{21}).$$

9.3.1 Without Masked Failure Modes

As mentioned in Section 9.2, in the case when the reliability testing is on device with competing failure modes, it is common for the experimenter to perform an autopsy on failed devices to determine the precise failure mode. It is possible, however, that during the autopsy process, the exact failure mode may not be determinable for some failed devices. In this case, we say that these devices have masked failure mode. In fact, when there are multiple (more than two) failure modes, though the exact failure mode may be unknown, it may be possible to narrow down the failure to a subset of all potential failure modes. This corresponds to masked failure modes in this general case; see Basu (2008), for example.

In this section, we assume that there are no masked failure modes in the data, indicating that the failure mode for each device is identified in the data. We then denote $p_{i,0}, p_{i,1}$, and $p_{i,2}$ for the reliability, the probability with failure mode 1, and the probability with failure mode 2, respectively, which are as follows:

$$p_{i,0} = (1 - F_{E,i,1}(\tau_i))(1 - F_{E,i,2}(\tau_i)) = \exp(-(\lambda_{i,1} + \lambda_{i,2})\tau_i), \tag{9.5}$$

$$p_{i,1} = \left(\frac{\lambda_{i,1}}{\lambda_{i,1} + \lambda_{i,2}}\right)(1 - \exp(-(\lambda_{i,1} + \lambda_{i,2})\tau_i)), \tag{9.6}$$

$$p_{i,2} = \left(\frac{\lambda_{i,2}}{\lambda_{i,1} + \lambda_{i,2}}\right)(1 - \exp(-(\lambda_{i,1} + \lambda_{i,2})\tau_i)). \tag{9.7}$$

Then, the likelihood function of θ_{ECR} for the given data is

$$L(\theta_{\mathrm{ECR}}) \propto \prod_{i=1}^{I} p_{i,0}^{n_{i,0}} p_{i,1}^{n_{i,1}} p_{i,2}^{n_{i,2}}, \tag{9.8}$$

where $K_i = n_{i,0} + n_{i,1} + n_{i,2}$.

In the example of electro-explosive devices, the true lifetimes of the devices are not observable. Let the lifetime $T_{i,g,k}^{(\Delta_{i,k})}$ be defined as

$$T_{i,g,k}^{(\Delta_{i,k})} = \begin{cases} T_{i,g,k} | \{\tau_i < \min(T_{i,1,k}, T_{i,2,k})\}, & \text{when } \Delta_{i,k} = 0, \\ T_{i,g,k} | \{T_{i,1,k} < \min(T_{i,2,k}, \tau_i)\}, & \text{when } \Delta_{i,k} = 1, \\ T_{i,g,k} | \{T_{i,2,k} < \min(T_{i,1,k}, \tau_i)\}, & \text{when } \Delta_{i,k} = 2. \end{cases} \tag{9.9}$$

Let us denote $\theta_{\mathrm{ECR}}^{(m)}$ for the current estimate of θ_{ECR} in the mth iteration. Then, the complete data log-likelihood is given by

$$\begin{aligned} l^{\mathrm{complete}}(\theta_{\mathrm{ECR}}) &= \sum_{i=1}^{I}\sum_{k=1}^{K_i} \ln\left(f_{E,i,1}(T_{i,1,k}^{(\delta_{i,k})})\right) + \ln\left(f_{E,i,2}(T_{i,2,k}^{(\delta_{i,k})})\right) \\ &= \sum_{i=1}^{I} l_i^{\mathrm{complete}}(\theta_{\mathrm{ECR}}), \end{aligned} \tag{9.10}$$

where

$$l_i^{\text{complete}}(\theta_{\text{ECR}}) = \sum_{k=1}^{K_i} \left\{ \ln(\lambda_{i,1}) - \lambda_{i,1} T_{i,1,k}^{(\delta_{i,k})} + \ln(\lambda_{i,2}) - \lambda_{i,2} T_{i,2,k}^{(\delta_{i,k})} \right\}$$
$$= K_i \sum_{g=1}^{2} \ln(\lambda_{i,g}) - \sum_{g=1}^{2} \lambda_{i,g} \sum_{k=1}^{K_i} T_{i,g,k}^{(\delta_{i,k})}. \tag{9.11}$$

In the E-step of the EM algorithm, we shall take the expected value of the missing data, given the observed data and the current parameter estimates, for imputing the missing data. It is given by

$$Q^{(m)} = Q\left(\theta_{\text{ECR}}, \theta_{\text{ECR}}^{(m)}\right) = \sum_{i=1}^{I} Q_i^{(m)}, \tag{9.12}$$

where

$$Q_i^{(m)} = Q_i\left(\theta_{\text{ECR}}, \theta_{\text{ECR}}^{(m)}\right) = K_i \sum_{g=1}^{2} \ln(\lambda_{i,g}) - \sum_{g=1}^{2} \lambda_{i,g} h_{i,g}\left(\theta_{\text{ECR}}^{(m)}\right) \tag{9.13}$$

and

$$h_{i,g}\left(\theta_{\text{ECR}}^{(m)}\right) = n_{i,0} A_{i,g,0}^{(m)} + n_{i,1} A_{i,g,1}^{(m)} + n_{i,2} A_{i,g,2}^{(m)} \tag{9.14}$$

is the conditional expectation of the failure time of the gth component, and $A_{i,g,u}^{(m)} = E[T_{i,g}^{(u)} | \mathbf{z}, \theta_{\text{ECR}}^{(m)}]$, for $i = 1, 2, \ldots, I, g = 1, 2, u = 0, 1, 2$, are the corresponding conditional expectations that are as presented in Table 9.2. The derivations of these expressions, as done by Balakrishnan et al. (2015), are presented in Appendix E.

Table 9.2 The conditional expectations for different cases (without masked data).

Case	No. of cases	Expression
$\delta_{i,k} = 0$	$n_{i,0}$	$A_{i,1,0}^{(m)} = \tau_i + \frac{1}{\lambda_{i,1}^{(m)}}$ $A_{i,2,0}^{(m)} = \tau_i + \frac{1}{\lambda_{i,2}^{(m)}}$
$\delta_{i,k} = 1$	$n_{i,1}$	$A_{i,1,1}^{(m)} = \frac{1}{\lambda_{i,1}^{(m)}+\lambda_{i,2}^{(m)}} - \frac{\tau_i \exp(-(\lambda_{i,1}^{(m)}+\lambda_{i,2}^{(m)})\tau_i)}{1-\exp(-(\lambda_{i,1}^{(m)}+\lambda_{i,2}^{(m)})\tau_i)}$ $A_{i,2,1}^{(m)} = \frac{1}{\lambda_{i,2}^{(m)}} + \frac{1}{\lambda_{i,1}^{(m)}+\lambda_{i,2}^{(m)}} - \frac{\tau_i \exp(-(\lambda_{i,1}^{(m)}+\lambda_{i,2}^{(m)})\tau_i)}{1-\exp(-(\lambda_{i,1}^{(m)}+\lambda_{i,2}^{(m)})\tau_i)}$
$\delta_{i,k} = 2$	$n_{i,2}$	$A_{i,1,2}^{(m)} = \frac{1}{\lambda_{i,1}^{(m)}} + \frac{1}{\lambda_{i,1}^{(m)}+\lambda_{i,2}^{(m)}} - \frac{\tau_i \exp(-(\lambda_{i,1}^{(m)}+\lambda_{i,2}^{(m)})\tau_i)}{1-\exp(-(\lambda_{i,1}^{(m)}+\lambda_{i,2}^{(m)})\tau_i)}$ $A_{i,2,2}^{(m)} = \frac{1}{\lambda_{i,1}^{(m)}+\lambda_{i,2}^{(m)}} - \frac{\tau_i \exp(-(\lambda_{i,1}^{(m)}+\lambda_{i,2}^{(m)})\tau_i)}{1-\exp(-(\lambda_{i,1}^{(m)}+\lambda_{i,2}^{(m)})\tau_i)}$

In the M-step, the objective function for maximizing the overall likelihood function $l^{\text{complete}}(\boldsymbol{\theta}_{\text{ECR}})$ will be the summation of the partial objective functions, $Q_i^{(m)}$, given in (9.13). Upon substituting the link functions for the failure rates specified in (9.4) and differentiating the objective function with respect to $\boldsymbol{\theta}_{\text{ECR}} = (e_{10}, e_{11}, e_{20}, e_{21})$, we obtain the first-order derivatives of $Q^{(m)}$ as

$$\frac{\partial Q^{(m)}}{\partial e_{g0}} = \sum_{i=1}^{I} \left(K_i - \lambda_{i,g} h_{i,g}\left(\boldsymbol{\theta}_{\text{ECR}}^{(m)}\right)\right),$$

$$\frac{\partial Q^{(m)}}{\partial e_{g1}} = \sum_{i=1}^{I} \left(K_i - \lambda_{i,g} h_{i,g}\left(\boldsymbol{\theta}_{\text{ECR}}^{(m)}\right)\right) x_i,$$

for $i = 1, 2, \ldots, I$. The gradient vector is then given by

$$\boldsymbol{G}\left(\boldsymbol{\theta}_{\text{ECR}}, \boldsymbol{\theta}_{\text{ECR}}^{(m)}\right) = \left[\frac{\partial Q^{(m)}}{\partial e_{10}}, \frac{\partial Q^{(m)}}{\partial e_{11}}, \frac{\partial Q^{(m)}}{\partial e_{20}}, \frac{\partial Q^{(m)}}{\partial e_{21}} \right]. \tag{9.15}$$

It is evident that the first-order derivatives involve nonlinear terms, which means we have to find the maximum likelihood estimates by employing numerical methods. Here, we adopt the one-step Newton–Raphson method to solve the system of equations. The method requires the Hessian matrix $\boldsymbol{H}$, with the second-order derivatives of $Q^{(m)}$, given by

$$\frac{\partial^2 Q^{(m)}}{\partial e_{g0}^2} = -\sum_{i=1}^{I} \lambda_{i,g} h_{i,g}\left(\boldsymbol{\theta}_{\text{ECR}}^{(m)}\right),$$

$$\frac{\partial^2 Q^{(m)}}{\partial e_{g1}^2} = -\sum_{i=1}^{I} x_i^2 \lambda_{i,g} h_{i,g}\left(\boldsymbol{\theta}_{\text{ECR}}^{(m)}\right),$$

$$\frac{\partial^2 Q^{(m)}}{\partial e_{g0} \partial e_{g1}} = -\sum_{i=1}^{I} x_i \lambda_{i,g} h_{i,g}\left(\boldsymbol{\theta}_{\text{ECR}}^{(m)}\right),$$

$$\frac{\partial^2 Q^{(m)}}{\partial e_{1p} \partial e_{2q}} = \frac{\partial^2 Q^{(m)}}{\partial e_{2p} \partial e_{1q}} = 0,$$

for $g = 1, 2$ and $p, q = 0, 1$. Since $\partial Q^{(m)}/\partial e_{1p}$ is free of e_{2q}, and $\partial Q^{(m)}/\partial e_{2p}$ is free of e_{1q}, we find $\partial^2 Q^{(m)}/\partial e_{1p}\partial e_{2q} = 0$. So, the Hessian matrix takes on the form

$$\boldsymbol{H}\left(\boldsymbol{\theta}_{\text{ECR}}, \boldsymbol{\theta}_{\text{ECR}}^{(m)}\right) = \begin{bmatrix} \frac{\partial^2 Q^{(m)}}{\partial e_{10}^2} & \frac{\partial^2 Q^{(m)}}{\partial e_{10}\partial e_{11}} & 0 & 0 \\ \frac{\partial^2 Q^{(m)}}{\partial e_{10}\partial e_{11}} & \frac{\partial^2 Q^{(m)}}{\partial e_{11}^2} & 0 & 0 \\ 0 & 0 & \frac{\partial^2 Q^{(m)}}{\partial e_{20}^2} & \frac{\partial^2 Q^{(m)}}{\partial e_{20}\partial e_{21}} \\ 0 & 0 & \frac{\partial^2 Q^{(m)}}{\partial e_{20}\partial e_{21}} & \frac{\partial^2 Q^{(m)}}{\partial e_{21}^2} \end{bmatrix}. \tag{9.16}$$

With the gradient vector and the Hessian matrix as presented in (9.15) and (9.16), respectively, we can maximize the objective function, $Q^{(m)}$, iteratively by updating the parameter in the $(m+1)$th step of the iterative process as

$$\boldsymbol{\theta}_{\mathrm{ECR}}^{(m+1)} = \boldsymbol{\theta}_{\mathrm{ECR}}^{(m)} - \left(\boldsymbol{H}\left(\boldsymbol{\theta}_{\mathrm{ECR}}^{(m)}, \boldsymbol{\theta}_{\mathrm{ECR}}^{(m)}\right)\right)^{-1} \boldsymbol{G}\left(\boldsymbol{\theta}_{\mathrm{ECR}}^{(m)}, \boldsymbol{\theta}_{\mathrm{ECR}}^{(m)}\right). \tag{9.17}$$

We will then go back to the E-step for another iteration. We will repeat the E- and the M-steps iteratively until convergence occurs to the desired level of accuracy.

In addition to the estimate of model parameters, we may be interested in estimating some useful lifetime quantities at normal operating condition x_0, such as the probability of failure with the gth failure mode, given failure by time t, i.e.

$$\hat{F}_g(t) = P(T_g < T_u | \{\min(T_g, T_u) < t\}) = \frac{\hat{\lambda}_{0,g}}{\hat{\lambda}_{0,1} + \hat{\lambda}_{0,2}},$$

mean time for the gth failure mode

$$\hat{\mu}_{E,g} = \frac{1}{\hat{\lambda}_{0,g}},$$

reliability at mission time t

$$\hat{R}_{\mathrm{ECR}}(t) = \exp(-(\hat{\lambda}_{0,1} + \hat{\lambda}_{0,2})t),$$

and mean lifetime of devices

$$\hat{\mu}_{\mathrm{ECR}} = \frac{1}{\hat{\lambda}_{0,1} + \hat{\lambda}_{0,2}},$$

where $\hat{\lambda}_{0,g} = \exp(\hat{e}_{g0} + \hat{e}_{g1}x_0)$, for $g = 1, 2$.

9.3.2 With Masked Failure Modes

In this section, we consider the situation when there are some failed devices for which the failure mode is masked. In this case, following the set-up of Balakrishnan et al. (2015), we use the indicator $\Delta_{i,k} = -1$ in the case of masked failure mode, and modify (9.1) accordingly as follows:

$$\Delta_{i,k} = \begin{cases} -1 & \text{for}\{\min(T_{i,1,k}, T_{i,2,k}) \le \tau_i\} \cap \{\text{masked}\}, \\ 0 & \text{for}\{\min(T_{i,1,k}, T_{i,2,k}) \ge \tau_i\}, \\ 1 & \text{for}\{T_{i,1,k} < \min(\tau_i, T_{i,2,k})\} \cap \{\text{not masked}\}, \\ 2 & \text{for}\{T_{i,2,k} < \min(\tau_i, T_{i,1,k})\} \cap \{\text{not masked}\}. \end{cases} \tag{9.18}$$

For simplicity, we may assume that the occurrence of masked failure modes is independent of the underlying unobserved failure modes responsible for the failure. Let q represent the probability of masked failure mode. We then have the following expressions, analogous to those in (9.5)–(9.7):

$$P(\Delta_{i,k} = -1) = (1 - p_{i,0})q, \tag{9.19}$$

$$P(\Delta_{i,k} = 0) = p_{i,0}, \tag{9.20}$$

$$P(\Delta_{i,k} = 1) = p_{i,1}(1 - q), \tag{9.21}$$

$$P(\Delta_{i,k} = 2) = p_{i,2}(1 - q). \tag{9.22}$$

Furthermore, we have

$$E[T_{i,g,k}|\{\text{conditions of } T_{i,1,k} \text{ and } T_{i,2,k}\} \cap \{\text{masked failure mode}\}]$$
$$= E[T_{i,g,k}|\{\text{conditions of } T_{i,1,k} \text{ and } T_{i,2,k}\}]. \quad (9.23)$$

Hence, the expectations of $T_{i,g,k}^{(\delta_{i,k})}$ when $\delta_{i,k} = 0, 1, 2$, with masked failure modes are the same as those without masked failure modes. Let us denote the number of devices failed with masking in the ith test group by m_i. Also, the sum of the number of failures with masked failure modes, failure mode 1 and failure mode 2, $m_i, n_{i,1}, n_{i,2}$, and the number of surviving devices $n_{i,0}$ will be equal to K_i, the number of devices inspected at time τ_i and the condition x_i; that is, $K_i = m_i + n_{i,0} + n_{i,1} + n_{i,2}$. Consequently, the likelihood function of θ_{ECR}, analogous to the one in (9.8), is given by

$$L_{\text{masked}}(\theta_{\text{ECR}}) = C \prod_{i=1}^{I} p_{i,0}^{n_{i,0}} p_{i,1}^{n_{i,1}} p_{i,2}^{n_{i,2}} (1 - p_{i,0})^{m_i} (1 - q)^{n_{i,1}+n_{i,2}} q^{m_i}. \quad (9.24)$$

It needs to be mentioned here that the EM algorithm has been employed successfully for estimating model parameters when there are masked failure modes under competing risk models; see Craiu and Duchesne (2004) and Park (2005). Here, from the results of Balakrishnan et al. (2015), we present the expectations of $T_{i,g,k}$ for all possible values of $\Delta_{i,k}$ in Table 9.3.

Table 9.3 The conditional expectations of missing data for different cases (with masked data).

Case	No. of cases	$E\left[T_{i,1}^{(\delta_{i,k})} \mid \mathbf{z}, \theta_{\text{ECR}}^{(m)}\right]$
$\delta_{i,k} = -1$	m_i	$\frac{1}{\lambda_{i,1}^{(m)}} - \frac{\tau_i \exp(-(\lambda_{i,1}^{(m)}+\lambda_{i,2}^{(m)})\tau_i)}{1-\exp(-(\lambda_{i,1}^{(m)}+\lambda_{i,2}^{(m)})\tau_i)}$
$\delta_{i,k} = 0$	$n_{i,0}$	$\tau_i + \frac{1}{\lambda_{i,1}^{(m)}}$
$\delta_{i,k} = 1$	$n_{i,1}$	$\frac{1}{\lambda_{i,1}^{(m)}+\lambda_{i,2}^{(m)}} - \frac{\tau_i \exp(-(\lambda_{i,1}^{(m)}+\lambda_{i,2}^{(m)})\tau_i)}{1-\exp(-(\lambda_{i,1}^{(m)}+\lambda_{i,2}^{(m)})\tau_i)}$
$\delta_{i,k} = 2$	$n_{i,2}$	$\frac{1}{\lambda_{i,1}^{(m)}} + \frac{1}{\lambda_{i,1}^{(m)}+\lambda_{i,2}^{(m)}} - \frac{\tau_i \exp(-(\lambda_{i,1}^{(m)}+\lambda_{i,2}^{(m)})\tau_i)}{1-\exp(-(\lambda_{i,1}^{(m)}+\lambda_{i,2}^{(m)})\tau_i)}$
Case	**No. of cases**	$E\left[T_{i,2}^{(\delta_{i,k})} \mid \mathbf{z}, \theta_{\text{ECR}}^{(m)}\right]$
$\delta_{i,k} = -1$	m_i	$\frac{1}{\lambda_{i,2}^{(m)}} - \frac{\tau_i \exp(-(\lambda_{i,1}^{(m)}+\lambda_{i,2}^{(m)})\tau_i)}{1-\exp(-(\lambda_{i,1}^{(m)}+\lambda_{i,2}^{(m)})\tau_i)}$
$\delta_{i,k} = 0$	$n_{i,0}$	$\tau_i + \frac{1}{\lambda_{i,2}^{(m)}}$
$\delta_{i,k} = 1$	$n_{i,1}$	$\frac{1}{\lambda_{i,2}^{(m)}} + \frac{1}{\lambda_{i,1}^{(m)}+\lambda_{i,2}^{(m)}} - \frac{\tau_i \exp(-(\lambda_{i,1}^{(m)}+\lambda_{i,2}^{(m)})\tau_i)}{1-\exp(-(\lambda_{i,1}^{(m)}+\lambda_{i,2}^{(m)})\tau_i)}$
$\delta_{i,k} = 2$	$n_{i,2}$	$\frac{1}{\lambda_{i,1}^{(m)}+\lambda_{i,2}^{(m)}} - \frac{\tau_i \exp(-(\lambda_{i,1}^{(m)}+\lambda_{i,2}^{(m)})\tau_i)}{1-\exp(-(\lambda_{i,1}^{(m)}+\lambda_{i,2}^{(m)})\tau_i)}$

Then, the complete data log-likelihood for the ith test group presented in (9.11) needs to be modified as

$$l_i^{\text{complete}}(\boldsymbol{\theta}_{\text{ECR}}) = K_i \sum_{g=1}^{2} \ln(\lambda_{i,g}) - \sum_{g=1}^{2} \lambda_{i,g} h_{i,g}^{*}(\boldsymbol{\theta}_{\text{ECR}}^{(m)}), \tag{9.25}$$

where

$$h_{i,g}^{*}(\boldsymbol{\theta}_{\text{ECR}}^{(m)}) = m_i E\left[T_{i,g}^{(-1)} | \mathbf{z}, \boldsymbol{\theta}_{\text{ECR}}^{(m)}\right] + n_{i,0} A_{i,g,0}^{(m)} + n_{i,1} A_{i,g,1}^{(m)} + n_{i,2} A_{i,g,2}^{(m)}, \tag{9.26}$$

with $A_{i,g,u}^{(m)} = E[T_{i,g}^{(u)} | \mathbf{z}, \boldsymbol{\theta}_{\text{ECR}}^{(m)}]$. We make use of the expressions in (9.25) and (9.26) and proceed as before with the maximization procedure. This way, the EM algorithm in Section 9.3.1 gets adapted to the situation when the available data involve masked failure modes.

9.4 Likelihood Estimation for Weibull Distribution

Here, we assume that the lifetime distribution, $T_{i,g,k}$, is Weibull with shape parameter $\eta_{i,g}$ and scale parameter $\beta_{i,g}$ and then present the EM algorithm for obtaining the MLEs of the model parameters for the competing risks model based on one-shot device testing data with two independent failure modes. This work was originally developed by Balakrishnan et al. (2016b).

The corresponding pdf and cdf are given by

$$f_{W,i,g}(t) = \left(\frac{\eta_{i,g}}{\beta_{i,g}}\right)\left(\frac{t}{\beta_{i,g}}\right)^{\eta_{i,g}-1} \exp\left(-\left(\frac{t}{\beta_{i,g}}\right)^{\eta_{i,g}}\right), \tag{9.27}$$

$$F_{W,i,g}(t) = 1 - \exp\left(-\left(\frac{t}{\beta_{i,g}}\right)^{\eta_{i,g}}\right), \quad t > 0, \beta_{i,g} > 0, \eta_{i,g} > 0, \tag{9.28}$$

for $g = 1, 2, i = 1, 2, \ldots, I$. Both $\beta_{i,g}$ and $\eta_{i,g}$ are related to temperature x_i in log-linear functions of the form

$$\eta_{i,g} = \exp(r_{g0} + r_{g1} x_i) \quad \text{and} \quad \beta_{i,g} = \exp(s_{g0} + s_{g1} x_i), \tag{9.29}$$

for $g = 1, 2$.

Instead of working with Weibull lifetime distribution, it is more convenient to work with the log-lifetime, $V_{i,g,k} = \ln(T_{i,g,k})$, which follows the extreme value distribution; see Section 2.3.3 for pertinent details. The corresponding pdf and cdf of the extreme value distribution are given by

$$f_{V,i,g}(v) = \frac{1}{\sigma_{i,g}} \exp\left(\frac{v - \mu_{i,g}}{\sigma_{i,g}}\right) \exp\left(-\exp\left(\frac{v - \mu_{i,g}}{\sigma_{i,g}}\right)\right), \tag{9.30}$$

$$F_{V,i,g}(v) = 1 - \exp\left(-\exp\left(\frac{v - \mu_{i,g}}{\sigma_{i,g}}\right)\right), \tag{9.31}$$

where $-\infty < v < \infty$, $\mu_{i,g} = \ln(\beta_{i,g})$ and $\sigma_{i,g} = 1/\eta_{i,g}$.

Furthermore, under temperature x_0, the mean time for the gth failure mode is given by

$$\mu_{W,g} = \beta_{0,g}\Gamma\left(1 + \frac{1}{\eta_{0,g}}\right), \quad g = 1, 2,$$

while the mean lifetime of device is given by

$$\mu_{\text{WCR}} = \int_0^{\infty} \exp\{-\exp(\xi_{0,1}(t)) - \exp(\xi_{0,2}(t))\}dt,$$

where $\xi_{0,g}(t) = \exp\left(\frac{\ln(t)-\mu_{0,g}}{\sigma_{0,g}}\right)$, $\mu_{0,g} = s_{g0} + s_{g1}x_0$ and $\sigma_{0,g} = \exp(-r_{g0} - r_{g1}x_0)$. In this setting, the new model parameter is

$$\boldsymbol{\theta}_{\text{WCR}} = (r_{10}, r_{11}, r_{20}, r_{21}, s_{10}, s_{11}, s_{20}, s_{21}).$$

Here, we assume that there are no masked failure modes in the data. Then, the reliability, the probability with failure mode 1, and the probability with failure mode 2 are given by

$$p_{i,0} = \prod_{g=1}^{2}(1 - F_{W,i,g}(\tau_i)) = \exp\left(-\sum_{g=1}^{2}\left(\frac{\tau_i}{\beta_{i,g}}\right)^{\eta_{i,g}}\right), \tag{9.32}$$

$$p_{i,1} = \int_0^{\tau_i} \exp\left(-\sum_{g=1}^{2}\left(\frac{t}{\beta_{i,g}}\right)^{\eta_{i,g}}\right)\left(\frac{\eta_{i,1}}{\beta_{i,1}}\right)\left(\frac{t}{\beta_{i,1}}\right)^{\eta_{i,1}-1} dt, \tag{9.33}$$

$$p_{i,2} = \int_0^{\tau_i} \exp\left(-\sum_{g=1}^{2}\left(\frac{t}{\beta_{i,g}}\right)^{\eta_{i,g}}\right)\left(\frac{\eta_{i,2}}{\beta_{i,2}}\right)\left(\frac{t}{\beta_{i,2}}\right)^{\eta_{i,2}-1} dt. \tag{9.34}$$

Then, the likelihood function of $\boldsymbol{\theta}_{\text{WCR}}$ for the given data is

$$L(\boldsymbol{\theta}_{\text{WCR}}) \propto \prod_{i=1}^{I} p_{i,0}^{n_{i,0}} p_{i,1}^{n_{i,1}} p_{i,2}^{n_{i,2}}, \tag{9.35}$$

where $K_i = n_{i,0} + n_{i,1} + n_{i,2}$.

Let us denote $\boldsymbol{\theta}_{\text{WCR}}^{(m)}$ for the current estimate of $\boldsymbol{\theta}_{\text{WCR}}$ in the mth iteration. Then, in the EM algorithm, the complete data log-likelihood is given by

$$\begin{aligned} l^{\text{complete}}(\boldsymbol{\theta}_{\text{WCR}}) &= \sum_{i=1}^{I}\sum_{k=1}^{K_i} \ln\left(f_{W,i,1}\left(T_{i,1,k}^{(\delta_{i,k})}\right)\right) + \ln\left(f_{W,i,2}\left(T_{i,2,k}^{(\delta_{i,k})}\right)\right) \\ &= \sum_{i=1}^{I} l_i^{\text{complete}}(\boldsymbol{\theta}_{\text{WCR}}), \end{aligned} \tag{9.36}$$

where

$$l_i^{\text{complete}}(\boldsymbol{\theta}_{\text{WCR}}) = \sum_{k=1}^{K_i}\sum_{g=1}^{2}\left\{ -\ln(\sigma_{i,g}) + \ln\left(\xi_{i,g,k}^{(\delta_{i,k})}\right) - \xi_{i,g,k}^{(\delta_{i,k})}\right\}$$

$$= -K_i \sum_{g=1}^{2} \ln(\sigma_{i,g}) + \sum_{k=1}^{K_i} \sum_{g=1}^{2} \ln\left(\xi_{i,g,k}^{(\delta_{i,k})}\right) - \sum_{k=1}^{K_i} \sum_{g=1}^{2} \xi_{i,g,k}^{(\delta_{i,k})}, \tag{9.37}$$

and $\xi_{i,g,k} = \exp\left(\frac{v_{i,g,k} - \mu_{i,g}}{\sigma_{i,g}}\right)$.

In the E-step of the EM algorithm, we shall take the expected value of the missing data, given the observed data and the current parameter estimates, for imputing the missing data. It is given by

$$\begin{aligned} Q^{(m)} &= Q\left(\boldsymbol{\theta}_{\text{WCR}}, \boldsymbol{\theta}_{\text{WCR}}^{(m)}\right) \\ &= -\sum_{i=1}^{I} \sum_{g=1}^{2} \left\{ K_i \ln\left(\sigma_{i,g}\right) - h_{i,g}^{(**)}\left(\boldsymbol{\theta}_{\text{WCR}}^{(m)}\right) + h_{i,g}^{(0*)}\left(\boldsymbol{\theta}_{\text{WCR}}^{(m)}\right) \right\}, \end{aligned} \tag{9.38}$$

where

$$h_{i,g}^{(**)}\left(\boldsymbol{\theta}_{\text{WCR}}^{(m)}\right) = \sum_{u=0}^{2} n_{i,u} E\left[\ln\left(\xi_{i,g}^{(u)}\right) | \mathbf{z}, \boldsymbol{\theta}_{\text{WCR}}^{(m)}\right],$$

$$h_{i,g}^{(0*)}\left(\boldsymbol{\theta}_{\text{WCR}}^{(m)}\right) = \sum_{u=0}^{2} n_{i,u} E\left[\xi_{i,g}^{(u)} | \mathbf{z}, \boldsymbol{\theta}_{\text{WCR}}^{(m)}\right],$$

for $i = 1, 2, \ldots, I$, $g = 1, 2$, and $u = 0, 1, 2$.

In the M-step, to maximize the overall likelihood function $l^{\text{complete}}(\boldsymbol{\theta}_{\text{WCR}})$, from (9.38), upon substituting the link functions for the shape and scale parameters specified in (9.29), and differentiating the objective function with respect to $\boldsymbol{\theta}_{\text{WCR}} = (r_{10}, r_{11}, s_{10}, s_{11}, r_{20}, r_{21}, s_{20}, s_{21})$, we obtain the first-order derivatives of $Q^{(m)}$ as

$$\frac{\partial Q^{(m)}}{\partial r_{g0}} = \sum_{i=1}^{I} \left(\frac{1}{\sigma_{i,g}}\right) (-K_i + h_{i,g}^{(0*)}(\boldsymbol{\theta}_{\text{WCR}}^{(m)})),$$

$$\frac{\partial Q^{(m)}}{\partial r_{g1}} = \sum_{i=1}^{I} \left(\frac{x_i}{\sigma_{i,g}}\right) (-K_i + h_{i,g}^{(0*)}(\boldsymbol{\theta}_{\text{WCR}}^{(m)})),$$

$$\frac{\partial Q^{(m)}}{\partial s_{g0}} = \sum_{i=1}^{I} (K_i + h_{i,g}^{(**)}(\boldsymbol{\theta}_{\text{WCR}}^{(m)}) - h_{i,g}^{(1*)}(\boldsymbol{\theta}_{\text{WCR}}^{(m)})),$$

$$\frac{\partial Q^{(m)}}{\partial s_{g1}} = \sum_{i=1}^{I} x_i (K_i + h_{i,g}^{(**)}(\boldsymbol{\theta}_{\text{WCR}}^{(m)}) - h_{i,g}^{(1*)}(\boldsymbol{\theta}_{\text{WCR}}^{(m)})),$$

where

$$h_{i,g}^{(1*)}(\boldsymbol{\theta}_{\text{WCR}}^{(m)}) = \sum_{u=0}^{2} n_{i,u} E[\xi_{i,g}^{(u)} \ln(\xi_{i,g}^{(u)}) | \mathbf{z}, \boldsymbol{\theta}_{\text{WCR}}^{(m)}].$$

The gradient vector is then given by

$$\boldsymbol{G}(\boldsymbol{\theta}_{\text{WCR}}, \boldsymbol{\theta}_{\text{WCR}}^{(m)})$$

$$= \left[\frac{\partial Q^{(m)}}{\partial r_{10}}, \frac{\partial Q^{(m)}}{\partial r_{11}}, \frac{\partial Q^{(m)}}{\partial s_{10}}, \frac{\partial Q^{(m)}}{\partial s_{11}}, \frac{\partial Q^{(m)}}{\partial r_{20}}, \frac{\partial Q^{(m)}}{\partial r_{21}}, \frac{\partial Q^{(m)}}{\partial s_{20}}, \frac{\partial Q^{(m)}}{\partial s_{21}}\right]. \tag{9.39}$$

Here, we adopt the one-step Newton–Raphson method to solve the system of equations. The method requires the Hessian matrix $\boldsymbol{H}$, with the second-order derivatives of $Q^{(m)}$, given by

$$\frac{\partial^2 Q^{(m)}}{\partial r_{g0}^2} = -\sum_{i=1}^{I}\left(\frac{h_{i,g}^{(0*)}(\boldsymbol{\theta}_{\mathrm{WCR}}^{(m)})}{\sigma_{i,g}^2}\right),$$

$$\frac{\partial^2 Q^{(m)}}{\partial r_{g0}\partial s_{g1}} = -\sum_{i=1}^{I}x_i\left(\frac{h_{i,g}^{(0*)}(\boldsymbol{\theta}_{\mathrm{WCR}}^{(m)})}{\sigma_{i,g}^2}\right),$$

$$\frac{\partial^2 Q^{(m)}}{\partial r_{g1}^2} = -\sum_{i=1}^{I}x_i^2\left(\frac{h_{i,g}^{(0*)}(\boldsymbol{\theta}_{\mathrm{WCR}}^{(m)})}{\sigma_{i,g}^2}\right),$$

$$\frac{\partial^2 Q^{(m)}}{\partial s_{g0}^2} = \sum_{i=1}^{I}(h_{i,g}^{(**)}(\boldsymbol{\theta}_{\mathrm{WCR}}^{(m)}) - h_{i,g}^{(1*)}(\boldsymbol{\theta}_{\mathrm{WCR}}^{(m)}) - h_{i,g}^{(2*)}(\boldsymbol{\theta}_{\mathrm{WCR}}^{(m)})),$$

$$\frac{\partial^2 Q^{(m)}}{\partial s_{g0}\partial s_{g1}} = \sum_{i=1}^{I}x_i(h_{i,g}^{(**)}(\boldsymbol{\theta}_{\mathrm{WCR}}^{(m)}) - h_{i,g}^{(1*)}(\boldsymbol{\theta}_{\mathrm{WCR}}^{(m)}) - h_{i,g}^{(2*)}(\boldsymbol{\theta}_{\mathrm{WCR}}^{(m)})),$$

$$\frac{\partial^2 Q^{(m)}}{\partial s_{g1}^2} = \sum_{i=1}^{I}x_i^2(h_{i,g}^{(**)}(\boldsymbol{\theta}_{\mathrm{WCR}}^{(m)}) - h_{i,g}^{(1*)}(\boldsymbol{\theta}_{\mathrm{WCR}}^{(m)}) - h_{i,g}^{(2*)}(\boldsymbol{\theta}_{\mathrm{WCR}}^{(m)})),$$

$$\frac{\partial^2 Q^{(m)}}{\partial r_{g0}\partial s_{g0}} = -\sum_{i=1}^{I}\left(\frac{K_i - h_{i,g}^{(0*)}(\boldsymbol{\theta}_{\mathrm{WCR}}^{(m)}) - h_{i,g}^{(1*)}(\boldsymbol{\theta}_{\mathrm{WCR}}^{(m)})}{\sigma_{i,g}}\right),$$

$$\frac{\partial^2 Q^{(m)}}{\partial r_{g0}\partial s_{g1}} = \frac{\partial^2 Q^{(m)}}{\partial r_{g1}\partial s_{g0}} = -\sum_{i=1}^{I}x_i\left(\frac{K_i - h_{i,g}^{(0*)}(\boldsymbol{\theta}_{\mathrm{WCR}}^{(m)}) - h_{i,g}^{(1*)}(\boldsymbol{\theta}_{\mathrm{WCR}}^{(m)})}{\sigma_{i,g}}\right),$$

$$\frac{\partial^2 Q^{(m)}}{\partial r_{g1}\partial s_{g1}} = -\sum_{i=1}^{I}x_i^2\left(\frac{K_i - h_{i,g}^{(0*)}(\boldsymbol{\theta}_{\mathrm{WCR}}^{(m)}) - h_{i,g}^{(1*)}(\boldsymbol{\theta}_{\mathrm{WCR}}^{(m)})}{\sigma_{i,g}}\right),$$

$$\frac{\partial^2 Q^{(m)}}{\partial r_{1p}\partial r_{2q}} = \frac{\partial^2 Q^{(m)}}{\partial r_{2p}\partial r_{1q}} = \frac{\partial^2 Q^{(m)}}{\partial s_{1p}\partial s_{2q}} = \frac{\partial^2 Q^{(m)}}{\partial s_{2p}\partial s_{1q}} = \frac{\partial^2 Q^{(m)}}{\partial r_{1p}\partial s_{2q}} = \frac{\partial^2 Q^{(m)}}{\partial r_{2p}\partial s_{1q}} = 0,$$

where

$$h_{i,g}^{(2*)}(\boldsymbol{\theta}_{\mathrm{WCR}}^{(m)}) = \sum_{u=0}^{2}n_{i,u}E[\xi_{i,g}^{(u)}(\ln(\xi_{i,g}^{(u)}))^2|\mathbf{z}, \boldsymbol{\theta}_{\mathrm{WCR}}^{(m)}].$$

So, the sub-Hessian matrix for the gth failure mode takes on the form

$$\boldsymbol{H}_g(\boldsymbol{\theta}_{\mathrm{WCR}}, \boldsymbol{\theta}_{\mathrm{WCR}}^{(m)}) = \begin{bmatrix} \frac{\partial^2 Q^{(m)}}{\partial r_{g0}^2} & \frac{\partial^2 Q^{(m)}}{\partial r_{g0}\partial r_{g1}} & \frac{\partial^2 Q^{(m)}}{\partial r_{g0}\partial s_{g0}} & \frac{\partial^2 Q^{(m)}}{\partial r_{g0}\partial s_{g1}} \\ \frac{\partial^2 Q^{(m)}}{\partial r_{g0}\partial r_{g1}} & \frac{\partial^2 Q^{(m)}}{\partial r_{g1}^2} & \frac{\partial^2 Q^{(m)}}{\partial r_{g1}\partial s_{g0}} & \frac{\partial^2 Q^{(m)}}{\partial r_{g1}\partial s_{g1}} \\ \frac{\partial^2 Q^{(m)}}{\partial r_{g0}\partial s_{g0}} & \frac{\partial^2 Q^{(m)}}{\partial r_{g1}\partial s_{g0}} & \frac{\partial^2 Q^{(m)}}{\partial s_{g0}^2} & \frac{\partial^2 Q^{(m)}}{\partial s_{g0}\partial s_{g1}} \\ \frac{\partial^2 Q^{(m)}}{\partial r_{g0}\partial s_{g1}} & \frac{\partial^2 Q^{(m)}}{\partial r_{g1}\partial s_{g1}} & \frac{\partial^2 Q^{(m)}}{\partial s_{g0}\partial s_{g1}} & \frac{\partial^2 Q^{(m)}}{\partial s_{g1}^2} \end{bmatrix}. \tag{9.40}$$

Finally, the Hessian matrix takes on the form

$$\boldsymbol{H}(\boldsymbol{\theta}_{\mathrm{WCR}}, \boldsymbol{\theta}_{\mathrm{WCR}}^{(m)}) = \begin{bmatrix} \boldsymbol{H}_1(\boldsymbol{\theta}_{\mathrm{WCR}}, \boldsymbol{\theta}_{\mathrm{WCR}}^{(m)}) & \mathbf{0} \\ \mathbf{0} & \boldsymbol{H}_2(\boldsymbol{\theta}_{\mathrm{WCR}}, \boldsymbol{\theta}_{\mathrm{WCR}}^{(m)}) \end{bmatrix}, \tag{9.41}$$

where $\mathbf{0}$ is a 4-by-4 zero matrix.

With the gradient vector and the Hessian matrix as presented in (9.39) and (9.41), respectively, we can maximize the objective function, $Q^{(m)}$, iteratively by updating the parameter in the $(m+1)$th step of the iterative process as

$$\boldsymbol{\theta}_{\mathrm{WCR}}^{(m+1)} = \boldsymbol{\theta}_{\mathrm{WCR}}^{(m)} - (\boldsymbol{H}(\boldsymbol{\theta}_{\mathrm{WCR}}^{(m)}, \boldsymbol{\theta}_{\mathrm{WCR}}^{(m)}))^{-1}\boldsymbol{G}(\boldsymbol{\theta}_{\mathrm{WCR}}^{(m)}, \boldsymbol{\theta}_{\mathrm{WCR}}^{(m)}). \tag{9.42}$$

From the M-step, expressions for 24 conditional expectations are required for setting up the E-step; however, these conditional expectations are difficult to implement. To simplify the process, we assume that the current estimate is close to the next updated estimate, and thus we have the following approximations of the conditional expectations:

$$E[\ln(\xi_{i,g}^{(u)})|\mathbf{z}, \boldsymbol{\theta}_{\mathrm{WCR}}^{(m)}] \approx M'_{i,g,u}(0), \tag{9.43}$$

$$E[\xi_{i,g}^{(u)}|\mathbf{z}, \boldsymbol{\theta}_{\mathrm{WCR}}^{(m)}] \approx M_{i,g,u}(1), \tag{9.44}$$

$$E[\xi_{i,g}^{(u)}\ln(\xi_{i,g}^{(u)})|\mathbf{z}, \boldsymbol{\theta}_{\mathrm{WCR}}^{(m)}] \approx M'_{i,g,u}(1), \tag{9.45}$$

$$E[\xi_{i,g}^{(u)}(\ln(\xi_{i,g}^{(u)}))^2|\mathbf{z}, \boldsymbol{\theta}_{\mathrm{WCR}}^{(m)}] \approx M''_{i,g,u}(1), \tag{9.46}$$

where $M_{i,g,u}(t)$ is the moment generating function of $\xi_{i,g}^{(u)}$. The explicit expressions of the required derivatives of the moment generating functions are presented in Table 9.4. Finally, we will repeat the E-step and the M-step until convergence occurs to the desired level of accuracy.

Similarly, we use the indicator $\Delta_{i,k} = -1$ when there are some failed devices for which the failure mode is masked. In addition, it is assumed that the occurrence of masked failure modes is independent of the unobserved failure modes responsible for the failure. In this case, we have

$$M_{i,g,-1}(t) = M_{i,g,1}(t)\frac{p_{i,1}}{p_{i,-1}} + M_{i,g,2}(t)\frac{p_{i,2}}{p_{i,-1}}, \tag{9.47}$$

Table 9.4 The explicit expectations of $M_{i,g,u}(t)$, $M'_{i,g,u}(t)$, and $M''_{i,g,u}(t)$ for different cases (without masking).

Notation	Expression
$\xi_{i,g}$	$\exp\left(\frac{\ln(\tau_i)-\mu_{i,g}}{\sigma_{i,g}}\right)$
$M'_{i,g,0}(0)$	$\ln(\xi_{i,g}) + \exp(\xi_{i,g})\Gamma(0,\xi_{i,g})$
$M_{i,g,0}(1)$	$\exp(\xi_{i,g})\Gamma(2,\xi_{i,g})$
$M'_{i,g,0}(1)$	$\xi_{i,g}\ln(\xi_{i,g}) + M'_{i,g,0}(0) + 1$
$M''_{i,g,0}(1)$	$(1+\xi_{i,g}+\exp(\xi_{i,g}))(\ln(\xi_{i,g}))^2 + 2M'_{i,g,0}(0)$ $+\exp(\xi_{i,g})\left\{\gamma^2 + \frac{\pi^2}{6} - 2\xi_{i,g\,3}F_3(1,1,1;2,2,2;-\xi_{i,g})\right\}$ $+2\exp(\xi_{i,g})\ln(\xi_{i,g})(\gamma+\Gamma(0,\xi_{i,g}))$
$M'_{i,1,1}(0)$	$\frac{1}{p_{i,1}}\int_0^{\xi_{i,1}} \ln(x)\exp\left\{-\exp\left(\frac{\mu_{i,1}-\mu_{i,2}}{\sigma_{i,2}}\right)x^{\frac{\sigma_{i,1}}{\sigma_{i,2}}} - x\right\}dx$
$M_{i,1,1}(1)$	$\frac{1}{p_{i,1}}\int_0^{\xi_{i,1}} x\exp\left\{-\exp\left(\frac{\mu_{i,1}-\mu_{i,2}}{\sigma_{i,2}}\right)x^{\frac{\sigma_{i,1}}{\sigma_{i,2}}} - x\right\}dx$
$M'_{i,1,1}(1)$	$\frac{1}{p_{i,1}}\int_0^{\xi_{i,1}} x\ln(x)\exp\left\{-\exp\left(\frac{\mu_{i,1}-\mu_{i,2}}{\sigma_{i,2}}\right)x^{\frac{\sigma_{i,1}}{\sigma_{i,2}}} - x\right\}dx$
$M''_{i,1,1}(1)$	$\frac{1}{p_{i,1}}\int_0^{\xi_{i,1}} x(\ln(x))^2\exp\left\{-\exp\left(\frac{\mu_{i,1}-\mu_{i,2}}{\sigma_{i,2}}\right)x^{\frac{\sigma_{i,1}}{\sigma_{i,2}}} - x\right\}dx$
$M'_{i,1,2}(0)$	$\frac{1}{p_{i,2}}\{-\gamma - p_{i,0}M'_{i,1,0}(0) - p_{i,1}M'_{i,1,1}(0)\}$
$M_{i,1,2}(1)$	$\frac{1}{p_{i,2}}\{1 - p_{i,0}M_{i,1,0}(1) - p_{i,1}M_{i,1,1}(1)\}$
$M'_{i,1,2}(1)$	$\frac{1}{p_{i,2}}\{1 - \gamma - p_{i,0}M'_{i,1,0}(1) - p_{i,1}M'_{i,1,1}(1)\}$
$M''_{i,1,2}(1)$	$\frac{1}{p_{i,2}}\left\{-2\gamma + \gamma^2 + \frac{\pi^2}{6} - p_{i,0}M''_{i,1,0}(1) - p_{i,1}M''_{i,1,1}(1)\right\}$
$M'_{i,2,2}(0)$	$\frac{1}{p_{i,2}}\int_0^{\xi_{i,2}} \ln(x)\exp\left\{-\exp\left(\frac{\mu_{i,2}-\mu_{i,1}}{\sigma_{i,1}}\right)x^{\frac{\sigma_{i,2}}{\sigma_{i,1}}} - x\right\}dx$
$M_{i,2,2}(1)$	$\frac{1}{p_{i,2}}\int_0^{\xi_{i,2}} x\exp\left\{-\exp\left(\frac{\mu_{i,2}-\mu_{i,1}}{\sigma_{i,1}}\right)x^{\frac{\sigma_{i,2}}{\sigma_{i,1}}} - x\right\}dx$
$M'_{i,2,2}(1)$	$\frac{1}{p_{i,2}}\int_0^{\xi_{i,2}} x\ln(x)\exp\left\{-\exp\left(\frac{\mu_{i,2}-\mu_{i,1}}{\sigma_{i,1}}\right)x^{\frac{\sigma_{i,2}}{\sigma_{i,1}}} - x\right\}dx$
$M'_{i,2,2}(0)$	$\frac{1}{p_{i,2}}\int_0^{\xi_{i,2}} x(\ln(x))^2\exp\left\{-\exp\left(\frac{\mu_{i,2}-\mu_{i,1}}{\sigma_{i,1}}\right)x^{\frac{\sigma_{i,2}}{\sigma_{i,1}}} - x\right\}dx$
$M'_{i,2,1}(0)$	$\frac{1}{p_{i,1}}\{-\gamma - p_{i,0}M'_{i,2,0}(0) - p_{i,2}M'_{i,2,2}(0)\}$
$M_{i,2,1}(1)$	$\frac{1}{p_{i,1}}\{1 - p_{i,0}M_{i,2,0}(1) - p_{i,2}M_{i,2,2}(1)\}$
$M'_{i,2,1}(1)$	$\frac{1}{p_{i,1}}\{1 - \gamma - p_{i,0}M'_{i,2,0}(1) - p_{i,2}M'_{i,2,2}(1)\}$
$M''_{i,2,1}(1)$	$\frac{1}{p_{i,1}}\left\{-2\gamma + \gamma^2 + \frac{\pi^2}{6} - p_{i,0}M''_{i,2,0}(1) - p_{i,2}M''_{i,2,2}(1)\right\}$

where $p_{i,-1} = 1 - p_{i,0} = p_{i,1} + p_{i,2}$. We can then obtain the conditional expectations in (9.43)–(9.46) for $\delta_{i,k} = -1$ in the E-step and proceed in the same way in the M-step.

9.5 Bayesian Estimation

The Bayesian approach detailed in Chapter 3 can be adopted for the situation of one-shot devices with competing failure modes as well. In this section, we describe the Bayesian estimation method for model parameters under the exponential lifetime distribution as developed by Balakrishnan et al. (2016a). We use the prior distributions of the parameters presented earlier in Chapter 3, namely, Laplace prior, normal prior with noninformative one for the variance, and Dirichlet prior as an extension of the beta prior mentioned there.

9.5.1 Without Masked Failure Modes

Let $\pi(\boldsymbol{\theta}_{\text{ECR}})$ be the joint prior density of model parameter $\boldsymbol{\theta}_{\text{ECR}} = (e_{10}, e_{11}, e_{20}, e_{21})$. Then, the joint posterior density of $\boldsymbol{\theta}_{\text{ECR}}$, given the observed data $\mathbf{z} = \{\tau_i, x_i, \delta_{i,k}, i = 1, 2, \ldots, I, k = 1, 2, \ldots, K_i\}$, is

$$\pi(\boldsymbol{\theta}_{\text{ECR}}|\mathbf{z}) = \frac{L(\boldsymbol{\theta}_{\text{ECR}}|\mathbf{z})\pi(\boldsymbol{\theta}_{\text{ECR}})}{\int_{\boldsymbol{\theta}_{\text{ECR}}\in\boldsymbol{\Theta}} L(\boldsymbol{\theta}_{\text{ECR}}|\mathbf{z})\pi(\boldsymbol{\theta}_{\text{ECR}})d\boldsymbol{\theta}_{\text{ECR}}}.$$

$$\propto \prod_{i=1}^{I} p_{i,0}^{n_{i,0}} p_{i,1}^{n_{i,1}} p_{i,2}^{n_{i,2}} \pi(\boldsymbol{\theta}_{\text{ECR}}). \tag{9.48}$$

Again, the denominator in (9.48) is usually not in closed-form. The Bayesian estimates of parameters of interest, such as model parameters, failure rate of the gth failure mode, $\lambda_{0,g}$, reliability at mission time t, $R_{\text{ECR}}(t)$, the probability of failure with the gth failure mode given that failure by time t, $F_g(t) = P(T_g < T_u|\{\min(T_g, T_u) < t\})$, mean time to the gth failure mode, $\mu_{E,g}$, and mean lifetime of a device, μ_{ECR}, under the normal operating condition x_0, can all be obtained by Markov Chain Monte Carlo sampling scheme via Metropolis–Hastings algorithm; see Chapter 3 for all pertinent details. Here again, after the "burn-in" and "reduction" stages, a sequence of posterior samples from the posterior distribution, $\{\boldsymbol{\theta}_{\text{ECR}}^{(b)}, b = 1, 2, \ldots, B\}$ can be generated and $\lambda_{0,g}^{(b)} = \exp(e_{g0}^{(b)} + e_{g1}^{(b)}x_0)$ can readily be obtained from them. The respective parameters of interest can then be estimated as follows:

Quantity	Bayesianestimate
$\lambda_{0,g}$	$\frac{1}{B}\sum_{b=1}^{B}\lambda_{0,g}^{(b)}$
$R_{\text{ECR}}(t)$	$\frac{1}{B}\sum_{b=1}^{B}\exp(-(\lambda_{0,1}^{(b)}+\lambda_{0,2}^{(b)})t)$
$F_g(t)$	$\frac{1}{B}\sum_{b=1}^{B}\lambda_{0,g}^{(b)}/(\lambda_{0,1}^{(b)}+\lambda_{0,2}^{(b)})$
$\mu_{E,g}$	$\frac{1}{B}\sum_{b=1}^{B}(\lambda_{0,g}^{(b)})^{-1}$
μ_{ECR}	$\frac{1}{B}\sum_{b=1}^{B}(\lambda_{0,1}^{(b)}+\lambda_{0,2}^{(b)})^{-1}$

Also, the mean lifetime is an useful quantity for finding the $100q_0\%$ quantile t_0 of a device with exponential lifetime, given simply by

$$P(T > t_0) = \exp(-\lambda t_0) \;\Rightarrow\; t_0 = -\ln(q_0)/\lambda.$$

Consequently, the Bayesian estimate of this quantile will be

$$-\ln(q_0)\frac{1}{B}\sum_{b=1}^{B}\frac{1}{\lambda^{(b)}},$$

which is a constant multiple of the Bayesian estimate of mean lifetime. So the mean square error of the estimate of quantile t_0 will simply be the multiplication of the mean square error of the estimate of mean lifetime $\frac{1}{\lambda}$ by $(\ln(q_0))^2$.

The following prior distributions can be used in developing the Bayesian inference described above:

9.5.2 Laplace Prior

Laplace prior is a simple prior distribution for model parameter $\boldsymbol{\theta}_{\text{ECR}} = (e_{10}, e_{11}, e_{20}, e_{21})$, which is of the form

$$\pi_L(\boldsymbol{\theta}_{\text{ECR}}) \propto \exp\left(-\sum_{g=1}^{2}\left|\frac{e_{g0}}{e_{g0}^h}\right| + \left|\frac{e_{g1}}{e_{g1}^h}\right|\right), \tag{9.49}$$

for g = 1, 2.

Then, from (9.48), the joint posterior density of $\boldsymbol{\theta}_{\text{ECR}}$ becomes

$$\pi_L(\boldsymbol{\theta}_{\text{ECR}}|\mathbf{z}) \propto \prod_{i=1}^{I} p_{i,0}^{n_{i,0}} p_{i,1}^{n_{i,1}} p_{i,2}^{n_{i,2}} \exp\left(-\sum_{g=1}^{2}\left|\frac{e_{g0}}{e_{g0}^h}\right| + \left|\frac{e_{g1}}{e_{g1}^h}\right|\right). \tag{9.50}$$

It is worth noting that e_{gj}^h is the unknown hyperparameter, which is usually based on experts' information in practice.

9.5.3 Normal Prior

Analogous to the normal prior described in Chapter 3, let $\epsilon_{i,g}$ be the error such that

$$p^h_{i,g} = p_{i,g} + \epsilon_{i,g}, \tag{9.51}$$

and let us now assume that the error $\epsilon_{i,g}$ are i.i.d. $N(0, \sigma^2)$ variables. Then, the conditional likelihood function of $\boldsymbol{\theta}_{\text{ECR}}$, given σ^2, is

$$L(\boldsymbol{\theta}_{\text{ECR}}|\mathbf{z}, p^h_{i,g}, \sigma^2) \propto \prod_{i=1}^{I}\prod_{g=1}^{2} \frac{1}{\sqrt{2\pi\sigma^2}} \exp\left\{-\frac{1}{2\sigma^2}(p_{i,g} - p^h_{i,g})^2\right\},$$

where $p_{i,1}$ and $p_{i,2}$ are as specified in (9.6) and (9.7), with $p_{i,g}$ denoting the true values, and $p^h_{i,g}$ is the unknown hyperparameter. We will now adopt the likelihood function as the prior distribution of $\boldsymbol{\theta}_{\text{ECR}}$:

$$\pi_N(\boldsymbol{\theta}_{\text{ECR}}|\mathbf{z}, \sigma^2) \propto \prod_{i=1}^{I}\prod_{g=1}^{2} \frac{1}{\sqrt{2\pi\sigma^2}} \exp\left\{-\frac{1}{2\sigma^2}(p_{i,g} - p^h_{i,g})^2\right\}. \tag{9.52}$$

As σ^2 is unknown, we adopt the noninformative prior

$$\pi(\sigma^2) \propto \frac{1}{\sigma^2}, \quad \sigma^2 > 0,$$

which yields the joint prior density of $\boldsymbol{\theta}_{\text{ECR}}$ as

$$\begin{aligned}
\pi_N(\boldsymbol{\theta}_{\text{ECR}}|\mathbf{z}) &\propto \int_0^\infty \pi_N(\boldsymbol{\theta}_{\text{ECR}}|\mathbf{z}, \sigma^2)\pi(\sigma^2)d\sigma^2 \\
&\propto \int_0^\infty (\sigma^2)^{-\frac{2I+2}{2}} \exp\left\{-\frac{1}{2\sigma^2}\sum_{i=1}^{I}\sum_{g=1}^{2}(p_{i,g} - p^h_{i,g})^2\right\} d\sigma^2 \\
&\propto \left\{\sum_{i=1}^{I}\sum_{g=1}^{2}(p_{i,g} - p^h_{i,g})^2\right\}^{-I}.
\end{aligned} \tag{9.53}$$

Then, from (9.48), the joint posterior density of $\boldsymbol{\theta}_{\text{ECR}}$ becomes

$$\pi_N(\boldsymbol{\theta}_{\text{ECR}}|\mathbf{z}) \propto \prod_{i=1}^{I} p_{i,0}^{n_{i,0}} p_{i,1}^{n_{i,1}} p_{i,2}^{n_{i,2}} \left\{\sum_{i=1}^{I}\sum_{g=1}^{2}(p_{i,g} - p^h_{i,g})^2\right\}^{-I}. \tag{9.54}$$

9.5.4 Dirichlet Prior

The natural extension of the beta prior described in Chapter 3 to the competing risks setup is the Dirichlet prior, with the density

$$f_i(p_{i,0}, p_{i,1}, p_{i,2}) = \frac{p_{i,0}^{\beta_{i0}-1} p_{i,1}^{\beta_{i1}-1} p_{i,2}^{\beta_{i2}-1}}{\Gamma_i},$$

where $p_{i,0} + p_{i,1} + p_{i,2} = 1$, $p_{i,0} > 0, p_{i,1} > 0, p_{i,2} > 0$, and

$$\Gamma_i = \frac{\Gamma(\beta_{i0})\Gamma(\beta_{i1})\Gamma(\beta_{i2})}{\Gamma(\beta_{i0} + \beta_{i1} + \beta_{i2})}.$$

The hyperparameters $(\beta_{i0}, \beta_{i1}, \beta_{i2})$ can then be chosen so as to match

$$\mathrm{E}[p_{i,g}] = \frac{\beta_{ig}}{\beta_{i0} + \beta_{i1} + \beta_{i2}} = p^h_{i,g}, \quad g = 0, 1, 2. \tag{9.55}$$

Clearly, one more issue needs to be taken care of in the determination of the hyperparameters. As mentioned earlier in Chapter 3, Fan et al. (2009) supposed that the prior belief of $p^h_{i,g}$ to be quite reliable with regard to the true unknown parameter $p_{i,g}$. So they generated $p_{i,g}$ from a beta distribution with specific choice of parameters. In the present case of competing risks in one-shot device testing data, we focus on the variance of $p_{i,0}$, corresponding to the accuracy of estimation of reliability of one-shot devices, and fix the prior belief such that $V[p_{i,0}] = c^2$. This results in the last equation for determining the hyperparameters as

$$V[p_{i,0}] = \frac{\beta_{i0}(\beta_{i1} + \beta_{i2})}{(\sum_{g=0}^{2}\beta_{ig})^2(\sum_{g=0}^{2}\beta_{ig} + 1)} = c^2. \tag{9.56}$$

Now upon using (9.55) and (9.56), we can choose the hyperparameters as

$$\beta_{i1} = p^h_{i,1}\left(\frac{p^h_{i,0}(1 - p^h_{i,0})}{c^2} - 1\right), \tag{9.57}$$

$$\beta_{i2} = p^h_{i,2}\left(\frac{p^h_{i,0}(1 - p^h_{i,0})}{c^2} - 1\right), \tag{9.58}$$

$$\beta_{i0} = \frac{p^h_{i,0}(1 - p^h_{i,0})}{c^2} - 1 - \beta_{i1} - \beta_{i2}, \tag{9.59}$$

yielding the corresponding posterior distribution as

$$\pi_D(\boldsymbol{\theta}_{\mathrm{ECR}}|\mathbf{z}) \propto \prod_{i=1}^{I} p_{i,0}^{n_{i,0}+\beta_{i0}-1} p_{i,1}^{n_{i,1}+\beta_{i1}-1} p_{i,2}^{n_{i,2}+\beta_{i2}-1}; \tag{9.60}$$

here, $p_{i,0}, p_{i,1}$ and $p_{i,2}$ are as specified earlier in (9.5)–(9.7), respectively. Note that $\sum_{g=0}^{2} \beta_{ig}$ has to be positive, implying that $c^2 < p^h_{i,0}(1 - p^h_{i,0})$.

9.5.5 With Masked Failure Modes

In this situation when the data contain some devices with masked failure modes, if the probability of masking is not of interest, then we may treat q to be an unknown constant and the posterior distribution in (9.24), with the prior distribution $\pi(\theta_{\text{ECR}})$, will be

$$\pi_{\text{masked}}(\theta_{\text{ECR}}|\mathbf{z}) = \frac{L_{\text{masked}}(\theta_{\text{ECR}}|\mathbf{z})\pi(\theta_{\text{ECR}})}{\int_{\theta_{\text{ECR}}\in\Theta} L_{\text{masked}}(\theta_{\text{ECR}}|\mathbf{z})\pi(\theta_{\text{ECR}})d\theta_{\text{ECR}}}$$

$$\propto \prod_{i=1}^{I} p_{i,0}^{n_{i,0}} p_{i,1}^{n_{i,1}} p_{i,2}^{n_{i,2}} (1 - p_{i,0})^{m_i} \pi(\theta_{\text{ECR}}). \tag{9.61}$$

The resultant posterior distribution in this case is similar to the original one presented in (9.48). It is then easy to work the posterior distributions for different prior distributions, such as Laplace, normal, and Dirichlet distributions, detailed in the preceding subsection.

9.6 Simulation Studies

In this section, we evaluate the performance of the EM algorithm and the Bayesian approach with various priors for one-shot device testing data with competing risks under exponential lifetime distribution. For this purpose, we consider CSALTs with $I = 12$ test groups consisting of three inspection times and four levels of temperatures, and then repeat this experiment with different sample sizes. For convenience, we allocate the same number of devices to each test group and denote the number by $K = K_i \in \{10, 30, 50\}$ for $i = 1, 2, \ldots, I$. The setting used in this simulation study are as given in Table 9.5.

Observe in Table 9.5 that failure mode 2 has a larger slope ($e_{21} = 0.08$) than failure mode 1 ($e_{11} = 0.05$). This means that failure mode 2 (the leakage of organic fuel) is more sensitive to the temperature than failure mode 1 (the burnout of resistance wire). This simulation setting, thus, imitates the higher chances of having cracks at higher temperature, while the resistance wire is not so sensitive to change in temperature. Moreover, we observe in Table 9.5 that failure mode 1 has a larger intercept ($e_{10} = -5.5$) than failure mode 2 ($e_{20} = -7.8$). This setting mimics the fact that most of the common failures of electro-explosive devices are due to the disconnection of resistance wires before ignition. The disconnection may be due to shocks (which is not the type of stress we are concerned with here) based on daily use, for example. The intercepts are small enough so that we have the devices to be of high reliability.

In this simulation study, 1000 sets of data were simulated under the specified settings. For the EM algorithm, the iteration was terminated when the norm of estimates of the model parameters (i.e. $||\theta_{\text{ECR}}^{(m)} - \theta_{\text{ECR}}^{(m+1)}||$) was less than 1×10^{-5}. For the Bayesian estimates, we used the Metropolis–Hastings algorithm (Hastings, 1970) to simulate the posterior distributions. We generated a vector of four normal random variables with mean parameters equal to the previous estimates, and

Table 9.5 Parameter values for exponential lifetime distribution used in the simulation of one-shot devices with two competing failure modes.

Parameters	Notation	Values
Failure mode 1	(e_{10}, e_{11})	(−5.5,0.05)
Failure mode 2	(e_{20}, e_{21})	(−7.8,0.08)
Temperature (°C)	x_i	{35,45,55,65}
Inspection time (days)	τ_i	{10,20,30}

the scale parameters as $(\sigma_0, \sigma_1) = (0.1, 0.001)$ for $(e_{g0}, e_{g1}), g = 1, 2$. We simulated a sequence of 100 000 values from the algorithm, and the first 1000 values were discarded as burn-in. We then chose one sample for every 100 values simulated to avoid correlation between the iterated samples; this way a sample of 990 observations was finally obtained. We also set $V[p_{i,0}] = c^2 = 0.005$ to avoid negative β values for Dirichlet prior. To evaluate the performance of these estimators, we then use their bias and root mean square error (RMSE), based on 1000 Monte Carlo simulations.

In addition, in order to evaluate the estimation methods for the competing risks model with masked data, we adopt the same settings as before and then take the probability of masked failure modes with $q = 0.3$. The bias and RMSE of the estimators are then determined based on 1000 simulations.

Tables 9.6 and 9.7 show the bias and RMSE of the estimates determined from the EM algorithm and the Bayesian approach with various priors under competing risks model without masked data. In general, both bias and RMSE decrease with K. In addition, we see that the EM algorithm performs better than the Bayesian approach with Laplace and Dirichlet priors. However, the EM algorithm becomes less efficient than the Bayesian approach with normal prior for the parameters of interest, namely, model parameters, $(e_{10}, e_{11}, e_{20}, e_{21})$, reliability at different mission times, $R_{ECR}(t)$, mean lifetime, μ_{ECR}, and the probability of failure with failure mode 1 given failure by time $t, F_1(t)$. Balakrishnan et al. (2015, 2016a) also compared the estimation methods at different levels of reliability and they observed the performance of the EM algorithm to be quite satisfactory for devices of low or moderate reliability. One reason for the inferior performance of the EM algorithm for devices of high reliability is that the MLEs depend solely on the observed data. Samples with devices of high reliability will have small number of observed failures and, therefore, the estimates will not be as accurate as those from samples with devices of moderate or low reliability. The corresponding results for masked data are presented in Tables 9.8 and 9.9. From these tables, we observe that results here are similar to those in Tables 9.6 and 9.7 when the data had no masking. But

Table 9.6 Bias of the estimates of parameters for devices with two competing failure modes under different estimation methods without masked data.

	Parameter value	e_{10} −5.5	e_{11} 0.05	e_{20} −7.8	e_{21} 0.08	$R_{ECR}(10)$ 0.841	$R_{ECR}(20)$ 0.708	$R_{ECR}(30)$ 0.595	μ_{ECR} 57.831	$F_1(10)$ 0.825
EM[a)]	$K = 10$	0.374	−0.010	0.412	−0.012	−0.030	−0.046	−0.054	−5.330	−0.022
	$K = 30$	0.102	−0.003	0.085	−0.003	−0.009	−0.015	−0.018	−1.670	−0.006
	$K = 50$	0.062	−0.002	0.017	−0.001	−0.005	−0.008	−0.010	−0.859	0.000
B.Lap[b)]	$K = 10$	0.268	−0.006	0.218	−0.005	−0.044	−0.065	−0.072	−3.896	−0.033
	$K = 30$	0.057	−0.001	0.025	−0.001	−0.012	−0.018	−0.021	−0.449	−0.012
	$K = 50$	0.045	−0.001	0.001	0.000	−0.008	−0.011	−0.013	−0.369	−0.005
B.Norm[c)]	$K = 10$	0.076	−0.002	0.128	−0.003	−0.015	−0.022	−0.024	0.417	−0.018
	$K = 30$	0.008	0.000	0.050	−0.001	−0.006	−0.008	−0.009	0.733	−0.012
	$K = 50$	0.005	0.000	−0.005	0.000	−0.003	−0.005	−0.005	0.713	−0.005
B.Dir[d)]	$K = 10$	0.403	−0.010	0.553	−0.014	−0.035	−0.055	−0.064	−5.443	−0.023
	$K = 30$	0.292	−0.007	0.387	−0.010	−0.023	−0.037	−0.044	−5.325	−0.014
	$K = 50$	0.223	−0.006	0.258	−0.007	−0.017	−0.027	−0.032	−3.925	−0.008

a) EM algorithm.
b) Bayesian with Laplace prior.
c) Bayesian with normal prior.
d) Bayesian with Dirichlet prior.

Table 9.7 RMSE of the estimates of parameters for devices with two competing failure modes under different estimation methods without masked data.

	Parameter value	e_{10} **−5.5**	e_{11} **0.05**	e_{20} **−7.8**	e_{21} **0.08**	$R_{ECR}(10)$ **0.8412**	$R_{ECR}(20)$ **0.7076**	$R_{ECR}(30)$ **0.5953**	μ_{ECR} **57.8309**	$F_1(10)$ **0.8249**
EM[a)]	$K = 10$	0.746	0.016	1.199	0.023	0.062	0.097	0.116	18.369	0.102
	$K = 30$	0.418	0.008	0.672	0.012	0.031	0.051	0.063	11.158	0.063
	$K = 50$	0.322	0.006	0.498	0.009	0.024	0.039	0.049	8.759	0.044
B.Lap[b)]	$K = 10$	0.770	0.015	1.193	0.022	0.073	0.111	0.128	20.089	0.101
	$K = 30$	0.438	0.009	0.689	0.013	0.033	0.054	0.066	11.937	0.064
	$K = 50$	0.330	0.007	0.515	0.009	0.025	0.041	0.051	9.129	0.044
B.Norm[c)]	$K = 10$	0.515	0.011	0.830	0.015	0.041	0.066	0.080	15.637	0.061
	$K = 30$	0.358	0.007	0.567	0.010	0.026	0.043	0.053	10.261	0.050
	$K = 50$	0.286	0.006	0.466	0.009	0.021	0.035	0.043	8.307	0.038
B.Dir[d)]	$K = 10$	0.818	0.019	1.229	0.026	0.059	0.094	0.112	42.706	0.074
	$K = 30$	0.518	0.012	0.808	0.017	0.040	0.065	0.078	11.945	0.058
	$K = 50$	0.424	0.010	0.636	0.013	0.032	0.052	0.063	10.151	0.043

a) EM algorithm.
b) Bayesian with Laplace prior.
c) Bayesian with normal prior.
d) Bayesian with Dirichlet prior.

Table 9.8 Bias of the estimates of parameters for devices with two competing failure modes under different estimation methods with masked data

	Parameter value	e_{10} −5.5	e_{11} 0.05	e_{20} −7.8	e_{21} 0.08	$R_{ECR}(10)$ 0.841	$R_{ECR}(20)$ 0.708	$R_{ECR}(30)$ 0.595	μ_{ECR} 57.831	$F_1(10)$ 0.825
EM[a]	$K = 10$	0.288	−0.009	−0.128	−0.004	−0.020	−0.031	−0.035	−1.554	−0.012
	$K = 30$	0.076	−0.002	0.096	−0.003	−0.009	−0.014	−0.016	−1.296	−0.011
	$K = 50$	0.056	−0.001	0.015	−0.001	−0.005	−0.008	−0.010	−0.819	−0.002
B.Lap[b]	$K = 10$	0.245	−0.005	0.202	−0.005	−0.046	−0.068	−0.076	−4.462	−0.045
	$K = 30$	0.041	−0.001	0.045	−0.001	−0.013	−0.019	−0.022	−0.617	−0.020
	$K = 50$	0.041	−0.001	−0.009	0.000	−0.008	−0.012	−0.014	−0.483	−0.008
B.Norm[c]	$K = 10$	0.077	−0.002	0.129	−0.003	−0.015	−0.022	−0.025	0.236	−0.020
	$K = 30$	0.002	0.000	0.064	−0.002	−0.006	−0.009	−0.009	0.651	−0.015
	$K = 50$	0.002	0.000	0.000	0.000	−0.003	−0.005	−0.005	0.656	−0.008
B.Dir[d]	$K = 10$	0.396	−0.010	0.567	−0.015	−0.035	−0.055	−0.065	−6.341	−0.025
	$K = 30$	0.278	−0.007	0.411	−0.011	−0.023	−0.037	−0.044	−5.341	−0.018
	$K = 50$	0.214	−0.006	0.271	−0.007	−0.017	−0.027	−0.032	−3.939	−0.011

a) EM algorithm.
b) Bayesian with Laplace prior.
c) Bayesian with normal prior.
d) Bayesian with Dirichlet prior.

Table 9.9 RMSE of the estimates of parameters for devices with two competing failure modes under different estimation methods with masked data.

	Parameter Value	e_{10} −5.5	e_{11} 0.05	e_{20} −7.8	e_{21} 0.08	$R_{ECR}(10)$ 0.8412	$R_{ECR}(20)$ 0.7076	$R_{ECR}(30)$ 0.5953	μ_{ECR} 57.8309	$F_1(10)$ 0.8249
EM[a)]	$K = 10$	0.765	0.016	5.468	0.085	0.059	0.094	0.113	20.347	0.125
	$K = 30$	0.436	0.009	0.763	0.014	0.031	0.051	0.064	11.385	0.075
	$K = 50$	0.341	0.007	0.562	0.010	0.024	0.040	0.050	8.926	0.053
B.Lap[b)]	$K = 10$	0.806	0.016	1.293	0.024	0.074	0.113	0.130	20.037	0.122
	$K = 30$	0.456	0.009	0.773	0.014	0.034	0.055	0.067	11.919	0.076
	$K = 50$	0.349	0.007	0.578	0.011	0.025	0.042	0.052	9.195	0.054
B.Norm[c)]	$K = 10$	0.511	0.011	0.858	0.016	0.041	0.066	0.080	15.375	0.064
	$K = 30$	0.361	0.007	0.604	0.011	0.026	0.043	0.053	10.204	0.055
	$K = 50$	0.293	0.006	0.501	0.009	0.021	0.035	0.044	8.378	0.043
B.Dir[d)]	$K = 10$	0.814	0.019	1.244	0.026	0.059	0.094	0.112	24.690	0.079
	$K = 30$	0.514	0.012	0.854	0.018	0.040	0.064	0.078	11.903	0.064
	$K = 50$	0.423	0.010	0.677	0.014	0.032	0.052	0.063	10.146	0.050

a) EM algorithm
b) Bayesian with Laplace prior
c) Bayesian with normal prior
d) Bayesian with Dirichlet prior

Table 9.10 Parameter values for Weibull lifetime distribution used in the simulation of one-shot devices with two competing failure modes.

Parameters	Notation	Values
Failure mode 1	$(r_{10}, r_{11}, s_{10}, s_{11})$	$(-1, 0.02, 6, -0.03)$
Failure mode 2	$(r_{20}, r_{21}, s_{20}, s_{21})$	$(-1.5, 0.03, 5.5, -0.04)$
Temperature (°C)	x_i	$\{35,45,55,65\}$
Inspection time (days)	τ_i	$\{10,20,30\}$

naturally, as there is more information missing in the case of masked data, we would expect the RMSE of the estimates based on masked data to be larger than those based on data without masking.

Furthermore, we evaluate the performance of the EM algorithm for one-shot device testing data with competing risks under Weibull lifetime distribution, and consider the same CSALTs for exponential lifetime distribution. The iteration was terminated when the norm of estimates of the model parameters (i.e. $||\theta_{WCR}^{(m)} - \theta_{WCR}^{(m+1)}||$) was less than 1×10^{-5}. The setting used in this simulation study are as given in Table 9.10.

Table 9.11 shows bias and RMSE of the estimates of the model parameters, reliability and mean lifetime, based on 1000 simulation experiments, determined from the EM algorithm for the competing risks model for Weibull lifetime distribution without masked data. In general, both bias and RMSE decrease with K. In addition, we see that the EM algorithm performs well when the sample size is greater than or equal to 50, except for the estimation of mean life. Compared to the results for exponential distribution without marking in Tables 9.8 and 9.9, we observe that the required sample size would be larger when the lifetime distribution is more complicated and flexible by incorporating more model parameters.

9.7 Case Study with R Codes

As an illustration, we apply the competing risks model and the methods to ED01 experiment data presented in Table 1.6. The R codes for defining the data with two competing risks are displayed in Table 9.12. It is noted that $x_i = 0$ for mice in a control group and $x_i = 1$ for mice injected with 150 ppm of a carcinogen. We treat the mice sacrificed at the inspection time as surviving mice, those died without

Table 9.11 Bias and RMSE of the estimates of parameters for devices with two competing failure modes for Weibull lifetime distribution without masking.

	r_{10}	r_{11}	s_{10}	s_{11}	r_{20}	r_{21}	s_{20}	s_{21}	$R_{WCR}(10)$	$R_{WCR}(20)$	$R_{WCR}(30)$	μ_{WCR}
Bias	**−1**	**0.02**	**6**	**−0.03**	**−1.5**	**0.03**	**5.5**	**−0.04**	**0.594**	**0.474**	**0.398**	**64.306**
$K = 50$	0.061	−0.002	0.336	0.000	−0.062	0.001	0.152	−0.002	−0.009	−0.016	−0.027	35.989
$K = 100$	0.074	−0.002	0.066	0.002	−0.001	0.000	0.043	−0.001	−0.001	−0.007	−0.013	3.815
$K = 200$	0.060	−0.002	−0.028	0.002	−0.002	0.000	0.029	0.000	0.000	−0.004	−0.008	0.109
RMSE	**−1**	**0.02**	**6**	**−0.03**	**−1.5**	**0.03**	**5.5**	**−0.04**	**0.594**	**0.474**	**0.398**	**64.306**
$K = 50$	1.141	0.021	3.401	0.055	0.684	0.012	0.778	0.013	0.074	0.063	0.070	336.846
$K = 100$	0.860	0.017	2.011	0.036	0.466	0.008	0.516	0.008	0.053	0.041	0.044	54.261
$K = 200$	0.670	0.013	1.348	0.025	0.335	0.006	0.358	0.006	0.040	0.029	0.029	25.969

Table 9.12 R codes for defining ED01 experiment data with two causes of death in Table 1.6 without and with masking.

```
#Without masked data
n0<-c(115,110,780,540,675,510)
n1<-c(22,49,42,54,200,64)
n2<-c(8,16,8,26,85,51)
n<-matrix(c(n0,n1,n2),ncol=3)
X<-c(0,1,0,1,0,1)
Tau<-c(12,12,18,18,33,33)
#With masked data
n0<-c(115,110,780,540,675,510)
n1<-c(21,38,34,48,173,55)
n2<-c(8,10,6,16,65,41)
m<-c(1,17,10,16,47,19)
n<-matrix(c(n0,n1,n2,m),ncol=4)
X<-c(0,1,0,1,0,1)
Tau<-c(12,12,18,18,33,33)
```

tumor as natural death, and those died with tumor as death due to cancer. The values of the hyperparameters chosen for Laplace, normal, and Dirichlet priors for exponential lifetime distribution are all listed in Table 9.13. Next, we modified the ED01 experimental data to introduce masked causes of death, and the data so obtained are presented in Table 9.14. Then, the R codes for the estimation of model parameters, mean lifetime, and survival probabilities at 10, 20, and 30 months for mice in the control group (not injected with the carcinogen) under the competing risks model without masking effect, by using the EM algorithm and the Bayesian approach with Laplace, normal, and Dirichlet priors are presented in Table 9.15. The corresponding outputs are presented in Table 9.17. From Table 9.17, we observe that the estimates of mean lifetime are sensitive to the choice of priors.

The R codes for defining the data and the estimation by the use of EM algorithm and the Bayesian approach with the three priors are displayed in Tables 9.12 and 9.16, respectively. Finally, the outputs for the masked data are presented in Table 9.17. We observe the results in Table 9.16 for the masked data to be similar to those when the data had no masking, except for the proportion of natural death among all the death cases in 10 months, $F_1(10)$.

Table 9.13 The values of the hyperparameters for Laplace, normal, and Dirichlet priors for exponential lifetime distribution for ED01 experiment data presented in Table 1.6.

Laplace	$(e^h_{10}, e^h_{11}, e^h_{20}, e^h_{21}) = (-5, -0.125, -6, 0.25)$
Dirichlet	$(p^h_{1,0}, p^h_{2,0}, p^h_{3,0}, p^h_{4,0}, p^h_{5,0}, p^h_{6,0}) = (0.9, 0.9, 0.86, 0.85, 0.75, 0.75)$
Normal/Dirichlet	$(p^h_{1,1}, p^h_{2,1}, p^h_{3,1}, p^h_{4,1}, p^h_{5,1}, p^h_{6,1}) = (0.07, 0.06, 0.1, 0.1, 0.18, 0.15)$
Normal/Dirichlet	$(p^h_{1,2}, p^h_{2,2}, p^h_{3,2}, p^h_{4,2}, p^h_{5,2}, p^h_{6,2}) = (0.03, 0.04, 0.04, 0.05, 0.07, 0.1)$

Table 9.14 The modified data from Table 1.6 with masked cause of death.

			Number of mice			
Test group	**Insp. time (months)**	**High dose of 2-AAF**	**Sacrificed**	**Died w/o tumor**	**Died w/ tumor**	**Died w/o cause**
1	12	No	115	21	8	1
2	12	Yes	110	38	10	17
3	18	No	780	34	6	10
4	18	Yes	540	48	16	16
5	33	No	675	173	65	47
6	33	Yes	510	55	41	19

Finally, we apply the EM algorithm to ED01 experiment data in Table 1.6 and the modified data with masked cause of death in Table 9.14 for the competing risks model under Weibull lifetime distribution. The R codes for the estimation of model parameters, mean lifetime, and survival probabilities for mice in the control group by using the EM algorithm are presented in Table 9.18. It needs to be mentioned that the R codes for Weibull lifetime distribution require the installation of several packages such as "gsl," "MASS," and "Rcpp" for incomplete gamma function, generalized inverse of a matrix and the integration of a C++ library "HyperGeom3F3fast.cpp" for the generalized hypergeometric function

Table 9.15 R codes for the estimation of model parameters, mean lifetime, and survival probabilities for mice in the control group for competing risks model under exponential lifetime distribution without masked data.

```
source("CR.R")
x0<-0 #control group
t0<-c(10,20,30)
#EM algorithm
K<-apply(n,1,sum)
p<-n0/K
p[p==0]<-0.01; p[p==1]<-0.99
y<-log(-log(1-p))-log(Tau)+log(n1)-log(n1+n2)
A<-matrix(c(rep(1,each=length(n0)),X),nrow=length(n0))
A<-A[is.finite(y),]
y<-y[is.finite(y)]
e1<-solve(t(A)%*%A,t(A)%*%y)
y<-log(-log(1-p))-log(Tau)+log(n2)-log(n1+n2)
A<-matrix(c(rep(1,each=length(n0)),X),nrow=length(n0))
A<-A[is.finite(y),]
y<-y[is.finite(y)]
e2<-solve(t(A)%*%A,t(A)%*%y)
theta0<-c(e1,e2)
EM<-EM.ECR(n,X,Tau,theta0,x0,t0,masked=F)
M<-1e6; D<-1e4; R<-1e3; sdp<-rep(1e-3,4)
h.L<-c(-5,-0.125,-6,0.25)
h.p0<-c(0.9,0.9,0.86,0.85,0.75,0.75)
h.p1<-c(0.07,0.06,0.1,0.1,0.18,0.15)
h.p2<-c(0.03,0.04,0.04,0.05,0.07,0.1)
h.N<-h.D<-matrix(c(h.p0,h.p1,h.p2),ncol=3)
Vp.L<-Vp.N<-NA; Vp.D<-0.005
hyper<-EM$theta.hat
#Bayesian approach with Laplace prior
BL<-Baye.ECR(n,X,Tau,h.L,x0,t0,M,"Lap",h.L,Vp.L,D,R,sdp,masked=F)
#Bayesian approach with normal prior
BN<-Baye.ECR(n,X,Tau,h.L,x0,t0,M,"norm",h.N,Vp.N,D,R,sdp,masked=F)
#Bayesian approach with Dirichlet prior
BD<-Baye.ECR(n,X,Tau,h.L,x0,t0,M,"Dir",h.D,Vp.D,D,R,sdp,masked=F)
```

Table 9.16 R codes for the estimation for competing risks model underexponential lifetime distribution with masked data.

```
#EM algorithm
K<-apply(n,1,sum)
p<-n0/K
p[p==0]<-0.01
p[p==1]<-0.99
y<-log(-log(1-p))-log(Tau)+log(n1)-log(n1+n2)
A<-matrix(c(rep(1,each=length(n0)),X),nrow=length(n0))
A<-A[is.finite(y),]
y<-y[is.finite(y)]
e1<-solve(t(A)%*%A,t(A)%*%y)
y<-log(-log(1-p))-log(Tau)+log(n2)-log(n1+n2)
A<-matrix(c(rep(1,each=length(n0)),X),nrow=length(n0))
A<-A[is.finite(y),]
y<-y[is.finite(y)]
e2<-solve(t(A)%*%A,t(A)%*%y)
theta0<-c(e1,e2)
EM<-EM.ECR(n,X,Tau,theta0,x0,t0,masked=T)
M<-1e6; D<-1e4; R<-1e3; sdp<-rep(1e-3,4)
h.L<-c(-5,-0.125,-6,0.25)
h.p0<-c(0.9,0.9,0.86,0.85,0.75,0.75)
h.p1<-c(0.07,0.06,0.1,0.1,0.18,0.15)
h.p2<-c(0.03,0.04,0.04,0.05,0.07,0.1)
h.N<-h.D<-matrix(c(h.p0,h.p1,h.p2),ncol=3)
Vp.L<-Vp.N<-NA; Vp.D<-0.005
hyper<-EM$theta.hat
#Bayesian approach with Laplace prior
BL<-Baye.ECR(n,X,Tau,h.L,x0,t0,M,"Lap",h.L,Vp.L,D,R,sdp,masked=T)
#Bayesian approach with normal prior
BN<-Baye.ECR(n,X,Tau,h.L,x0,t0,M,"norm",h.N,Vp.N,D,R,sdp,masked=T)
#Bayesian approach with Dirichlet prior
BD<-Baye.ECR(n,X,Tau,h.L,x0,t0,M,"Dir",h.D,Vp.D,D,R,sdp,masked=T)
```

Table 9.17 R outputs for ED01 experiment data in Table 1.6 and 9.14 for competing risks model under exponential lifetime distribution.

	No masked data in Table 1.6				Masked data in Table 9.14			
		Bayesian				Bayesian		
Parameter	MLE	Laplace[a)]	normal[b)]	Dirichlet[c)]	MLE	Laplace[a)]	normal[b)]	Dirichlet[c)]
e_{10}	−5.088[d)]	−5.113	−5.081	−5.084	−5.062[d)]	−5.083	−5.083	5.064
e_{11}	−0.128[d)]	−0.075	−0.144	−0.139	−0.100[d)]	−0.048	−0.134	−0.122
e_{20}	−6.049[d)]	−5.987	−6.010	−6.122	−6.121[d)]	−6.102	−6.037	−6.092
e_{21}	0.247[d)]	0.117	0.289	0.290	0.215[d)]	0.144	0.308	0.117
$R_{ECR}(10)$	0.918[e)]	0.918	0.917	0.919	0.918[e)]	0.919	0.918	0.918
$R_{ECR}(20)$	0.843[e)]	0.843	0.841	0.845	0.843[e)]	0.844	0.842	0.842
$R_{ECR}(30)$	0.774[e)]	0.774	0.771	0.777	0.774[e)]	0.776	0.773	0.773
μ_{ECR}	117.236[e)]	117.358	115.523	119.272	117.236[e)]	118.506	116.369	116.491
$F_1(10)$	0.723[e)]	0.705	0.717	0.738	0.743[e)]	0.734	0.722	0.736

a) EM$theta.hat;
b) EM$phi.hat;
c) BL$phi.hat;
d) BN$phi.hat;
e) BD$phi.hat

Table 9.18 R codes for the estimation for competing risks model under Weibull lifetime distribution.

```
source("CR.R")
x0<-0 #control group
t0<-c(10,20,30)
theta0<-c(0,0,-EM$theta.hat[1:2],0,0,-EM$theta.hat[3:4])
EMW<-EM.WCR(n,X,Tau,theta0,x0,t0,masked=F) #without masking
EMW<-EM.WCR(n,X,Tau,theta0,x0,t0,masked=T) #with masking
```

Table 9.19 R outputs for ED01 experiment data in Table 1.6 and the modified data with masked cause of death in Table 9.14 for competing risks model under Weibull lifetime distribution.

	No masked data in Table 1.6	Masked data in Table 9.14
r_{10} [a]	0.492	0.536
r_{11} [a]	−3.457	−3.581
s_{10} [a]	4.392	4.329
s_{11} [a]	38.836	40.980
r_{20} [a]	0.864	0.777
r_{21} [a]	−1.876	−1.563
s_{20} [a]	4.460	4.594
s_{21} [a]	5.772	4.442
$R_{WCR}(10)$ [b]	0.962	0.963
$R_{WCR}(20)$ [b]	0.875	0.875
$R_{WCR}(30)$ [b]	0.757	0.756
μ_{WCR} [b]	51.867	52.905

a) EMW$theta.hat;
b) EMW$phi.hat

${}_3F_3(a_1, a_2, a_3; b_1, b_2, b_3; z)$ in Table 9.4. Table 9.19 presents the corresponding output. We observe that the estimates of model parameters, reliability, and mean lifetime are quite similar to the estimates obtained from data without masked cause of death.

10

One-Shot Devices with Dependent Components

10.1 Brief Overview

Competing risks models have been discussed in Chapter 9, in which it was assumed that times to failure modes are independent. In this chapter, the independence between failure modes is relaxed and copula models are then used to model dependent failure modes in one-shot devices under constant-stress accelerated life-tests (CSALTs). A composite likelihood approach, as developed by Ling et al. (2020), is described here in detail for the likelihood estimation of model parameters with the use of two prominent Archimedean copulas, namely, Gumbel–Hougaard (GH) and Frank copulas.

10.2 Test Data with Dependent Components

For convenience, we focus here on one-shot devices with only two failure modes under CSALTs with I higher-than-normal stress conditions, each of which is subject to an accelerating factor, and with J inspection times. For $i = 1, 2, \dots, I$, and $j = 1, 2, \dots, J$, $K_{i,j}$ devices are placed at stress level x_i and get inspected at inspection time τ_j. The numbers of devices with no failure ($n_{i,j,0}$), with only failure mode 1 ($n_{i,j,1}$), with only failure mode 2 ($n_{i,j,2}$), and with both failure modes ($n_{i,j,12}$) are then recorded. The one-shot device testing data collected in such a manner may be summarized as in Table 10.1. For notational convenience, $\mathbf{z} = \{x_i, \tau_j, K_{i,j}, n_{i,j,0}, n_{i,j,1}, n_{i,j,2}, i = 1, 2, \dots, I, j = 1, 2, \dots, J\}$ is used to denote the observed data. Bear in mind that each one-shot device can be tested only once. Note that both failure modes may be observed at inspection times on some devices under this setting. Clearly, we have $K_{i,j} = n_{i,j,0} + n_{i,j,1} + n_{i,j,2} + n_{i,j,12}$, for $i = 1, 2, \dots, I, j = 1, 2, \dots, J$. We further denote the total number of devices with the gth failure mode by $N_{i,j,g} = n_{i,j,g} + n_{i,j,12}$, for $g = 1, 2$.

Accelerated Life Testing of One-shot Devices: Data Collection and Analysis, First Edition.
Narayanaswamy Balakrishnan, Man Ho Ling, and Hon Yiu So.

Companion Website: www.wiley.com/go/Balakrishnan/Accelerated_Life_Testing

Table 10.1 Form of one-shot device testing data with two failure modes.

	Stress level x_i, inspection time τ_j		
	Failure mode 1		
Failure mode 2	Present	Absent	Total
Present	$n_{i,j}, 12$	$n_{i,j}, 2$	$N_{i,j}, 12$
Absent	$n_{i,j}, 1$	$n_{i,j}, 0$	$K_{i,j}-N_{i,j}, 2$
Total	$N_{i,j}, 1$	$K_{i,j}-N_{i,j}, 1$	$K_{i,j}$

10.3 Copula Models

An independence assumption was made on the failure modes while discussing the analysis of one-shot device testing data in the presence of competing risks for failure in the Chapter 9. Even though this assumption is convenient and makes the subsequent analysis somewhat simpler, it may be unrealistic in many situations and consequently may have serious impact on the evaluation of reliability of devices. Sklar (1959), in his pioneering work, introduced copulas as a convenient way to model multivariate distributions. Copula models have recently become quite popular in reliability studies for capturing dependence between lifetime variables corresponding to different parts or components. For example, Pan et al. (2013) and Peng et al. (2016) used copula models, together with different stochastic processes, for analyzing products with two correlated degradation paths. Similarly, Hong et al. (2014) modeled systems with correlated degradation of components by copula models for condition-based maintenance decisions, while Jia et al. (2014) used copula models for measuring reliability of systems with dependent components.

In this chapter, we focus on two prominent special cases within the family of Archimedean copulas – Gumbel–Hougaard copula and Frank copula – for the analysis of data on ones-shot devices with two dependent failure modes. This discussion is based on the recent work of Ling et al. (2020).

It is important to mention here that, though we focus here only on Gumbel–Hougaard and Frank copulas, many more copulas with varying characteristics and properties are available in the statistical literature. Interested readers may refer to the concise description of copulas given by Kotz et al. (2000) and Balakrishnan and Lai (2010), or detailed booklength accounts on copulas provided by Nelsen (2007) and Joe (2014).

10.3.1 Family of Archimedean Copulas

For this discussion, let us assume (X, Y) corresponds to continuous lifetimes of two components within a reliability system, having marginal cumulative distributions as $F_X(x)$ and $F_Y(y)$, for $x, y \geq 0$. Then, it is well known that the probability integral transformations

$$U = F_X(X) \quad \text{and} \quad V = F_Y(Y)$$

will result in U and V being standard uniform random variables, but bring possibly dependent due to dependence between X and Y. Hence, with the joint distribution function of (U, V) being represented by $C(u, v)$, for $0 \leq u, v \leq 1$, and the marginal distributions $F_X(x)$ and $F_Y(y)$, Sklar's [1959] theorem facilitates splitting the joint distribution of (X, Y) into a "coupla function" C and the marginal distributions F_X and F_Y; see also Sklar (1973) and Schweizer (1991). Evidently, any choice of C needs to satisfy some monotonicity conditions in $[0, 1] \times [0, 1]$ as it represents the joint cumulative distribution function of (U, V). Thus, we have the representation for the joint lifetime distribution function of (X, Y) as

$$F_{X,Y}(x, y) = C(F_X(x), F_Y(y)), \quad x, y \geq 0.$$

The case of independence will correspond to the case of "product copula" or "independence copula" with

$$F_{X,Y}(x, y) = C(F_X(x), F_Y(y)) = F_X(x)F_Y(y), \quad x, y \geq 0.$$

Now, let $\phi(u)$ be a continuously decreasing convex function such that $\phi(1) = 0$. In the context of copulas, ϕ is called a "generator," and it is said to be a "strict generator" if $\phi(0) = +\infty$. Then, with ϕ being a strict generator and ϕ^{-1} being completely monotonic on $[0, \infty) \times [0, \infty)$, the family of Archimedean copulas is given by

$$C(u, v) = \phi^{-1}(\phi(u) + \phi(v)), \quad 0 \leq u, v \leq 1. \tag{10.1}$$

Any choice of ϕ possessing the abovementioned properties will result in a member of the family Archimedean copulas. Of course, one convenient choice for ϕ is through inverse of Laplace transforms of cumulative distribution functions as it is known that a function on $[0, \infty)$ is the Laplace transform of a cumulative distribution function F if and only if it is completely monotonic and is 1 at 0; see for example, Feller (1971).

An interesting and attractive property of Archimedean copulas in (10.1) is its simple and explicit connection to dependence measures such as Kendall's tau (Kendall 1938). It is defined as the difference between probabilities of concordance and discordance between two random variables. For the case of Archimedean copulas in (10.1); Genest and MacKay (1986) proved that Kendall's tau measure is simply (also see Cherubini et al. 2004; Huard et al. 2006; Nelsen 2007)

$$\tau = 1 + 4\int_0^1 \frac{\phi(t)}{\phi'(t)}dt. \tag{10.2}$$

10.3.2 Gumbel–Hougaard Copula

The Gumbel–Hougaard copula, a member of the Archimedean copula family in (10.1) that can characterize positive dependence between random variables, is a popular copula model which has found a wide range of applications; see Tovar and Achcar (2013); Zhang et al. (2014); Fang et al. (2020), and Barmalzan et al. (2020). This is the case when the generator function $\phi(t) = (-\ln t)^{\kappa}$ for $\kappa \geq 1$. The joint cumulative distribution function in this case is then obtained from (10.1) to be

$$C_{\kappa}(u, v) = P(U \leq u, V \leq v) = \exp(-\{(-\ln u)^{\kappa} + (-\ln v)^{\kappa}\}^{1/\kappa}), \tag{10.3}$$

where $\kappa \geq 1$, and U and V are standard uniformly distributed random variables; moreover, we have

$$\begin{aligned}
P(U \leq u, V > v) &= P(U \leq u) - P(U \leq u, V \leq v) = P(U \leq u) - C_{\kappa}(u, v),\\
P(U > u, V \leq v) &= P(V \leq v) - P(U \leq u, V \leq v) = P(V \leq v) - C_{\kappa}(u, v),\\
P(U > u, V > v) &= 1 - P(U \leq u) - P(V \leq v) + P(U \leq u, V \leq v)\\
&= 1 - P(U \leq u) - P(V \leq v) + C_{\kappa}(u, v).
\end{aligned}$$

Moreover, from (10.2), it can be shown that Kendall's tau measure in this case is simply

$$\tau = 1 - \frac{1}{\kappa}. \tag{10.4}$$

Now, returning to the problem of one-shot device testing data with two failure modes, let $T_{i,g}$ denote the time for the gth failure mode at stress level x_i, for $i = 1, 2, \ldots, I, g = 1, 2$. Further, let $T_{i,g}$ have its marginal distribution function as $F_{i,g}(t)$, for $g = 1, 2$. In order to introduce dependence between the two failure modes, we shall now assume that the joint distribution function of the lifetimes corresponding to the two failure modes is given by the Gumbel–Hougaard copula in (10.3) with u and v being the respective marginal distributions $F_{i,1}(t_1)$ and $F_{i,2}(t_2)$. Then, to allow the association between the lifetimes of the two failure modes to vary over stress levels, we take $\ln(\kappa_i - 1) = \ln\left(\frac{\tau_i}{1-\tau_i}\right)$ to relate to stress level x_i in a linear form as

$$\ln(\kappa_i - 1) = c_0 + c_1 x_i, \quad i = 1, 2, \ldots, I. \tag{10.5}$$

Then, the joint cumulative distribution function of lifetimes $T_{i,1}$ and $T_{i,2}$, under the considered copula can be expressed from (10.3) as

$$C_{\kappa_i}(F_{i,1}(t_1), F_{i,2}(t_2))$$

$$= P(T_{i,1} \leq t_1, T_{i,2} \leq t_2)$$
$$= \exp(-\{(-\ln F_{i,1}(t_1))^{\kappa_i} + (-\ln F_{i,2}(t_2))^{\kappa_i}\}^{1/\kappa_i}). \quad (10.6)$$

It is readily seen from (10.6) that $C_{\kappa_i}(F_{i,1}(t_1), F_{i,2}(t_2)) = F_{i,1}(t_1)F_{i,2}(t_2)$ when $\kappa_i = 1$, which implies that $T_{i,1}$ and $T_{i,2}$ are independent at stress level x_i. Moreover, the dependence parameter remains unchanged over stress levels when $c_1 = 0$.

As the occurrence of any one of the failure modes results in the failure of the device, the lifetime variable of the device at stress level x_i is given by $T_i = \min\{T_{i,1}, T_{i,2}\}$, and so the reliability of the device at mission time t is then

$$R_i(t) = P(T_{i,1} > t, T_{i,2} > t) = 1 - F_{i,1}(t) - F_{i,2}(t) + C_{\kappa_i}(F_{i,1}(t), F_{i,2}(t)). \quad (10.7)$$

In the special case when $\kappa_i = 1$, we deduce that

$$R_i(t) = 1 - F_{i,1}(t) - F_{i,2}(t) + F_{i,1}(t)F_{i,2}(t)$$
$$= (1 - F_{i,1}(t))(1 - F_{i,2}(t)) = R_{i,1}(t)R_{i,2}(t). \quad (10.8)$$

However, $T_{i,1}$ and $T_{i,2}$ become highly positively associated when κ_i tends to ∞. Also, if $T_{i,1}$ is less than $T_{i,2}$ in the stochastic order, i.e. $R_{i,1}(t) \leq R_{i,2}(t)$ for all $t > 0$, we will get in the limit (as association becomes stronger)

$$R_i(t) = R_{i,1}(t) - F_{i,2}(t) + \lim_{\kappa_i \to \infty} C_{\kappa_i}(F_{i,1}(t), F_{i,2}(t)) = R_{i,1}(t). \quad (10.9)$$

This is the upper Fréchet–Hoeffding copula bound; see Fréchet (1957) and Kotz et al. (2000).

It is evident from (10.9) that, when failure mode 1 is more likely to occur than failure mode 2 at every possible inspection time, then as the failure modes become highly positively associated, then the reliability of the device will simply become the reliability corresponding to failure mode 1, agreeing with our intuition. Further, the reliability of the device will become $R_1(t)R_2(t)$ when the lifetimes corresponding to the two failure modes are independent. This also reveals that devices with two independent failure modes would have shorter lifetimes than those with highly positively associated failure modes, which makes sense in that a longer time for one failure mode to occur does not provide any information on the time for the other failure mode to occur when the times to the two failure modes are independent.

10.3.3 Frank Copula

The Frank copula is another commonly used copula model within the family of Archimedean copulas, with its generator function as $\phi(t) = \ln(\exp(-\kappa) - 1) - \ln(\exp(-\kappa t) - 1)$. This copula has also been used extensively in different applications; see, for example, Chiyoshi (2018); Pan et al. (2013); Fang et al.

(2020), and Barmalzan et al. (2020). The joint cumulative distribution function in this case is obtained from (10.1) to be

$$C_\kappa(u, v) = -\frac{1}{\kappa} \ln \left\{ 1 + \frac{(\exp(-\kappa u) - 1)(\exp(-\kappa v) - 1)}{\exp(-\kappa) - 1} \right\}, \tag{10.10}$$

for $\kappa \in (-\infty, \infty)\backslash\{0\}$; otherwise, $C_\kappa(u, v) = uv$, where U and V are again standard uniformly distributed random variables. This copula allows the modeling of positive ($\kappa > 0$) and negative ($\kappa < 0$) dependence between the two random variables and is, therefore, more flexible. The independence case is attained when κ approaches zero.

For Frank copula, the Archimedean generator function, as mentioned above, is

$$\phi(t) = -\ln \left(\frac{\exp(-\kappa t) - 1}{\exp(-\kappa) - 1} \right), \tag{10.11}$$

and when $\kappa = 0$, we take the limit to get $\phi(t) = -\ln t$.

Kendall's tau measure in this case is

$$\tau = 1 - \frac{4}{\kappa}(1 - D_1(\kappa)), \tag{10.12}$$

where $D_n(x)$ is called Debye function defined by

$$D_n(x) = \frac{n}{x^n} \int_0^x \frac{t^n}{\exp(t) - 1} dt. \tag{10.13}$$

Of course, for the special case of $D_1(\kappa)$ involved in tau measure in (10.12), we have a simple expression

$$D_1(\kappa) = \frac{1}{\kappa} \int_0^\kappa \frac{t}{\exp(t) - 1} dt. \tag{10.14}$$

From the limiting values and expansions of Debye functions presented, for example, in Abramowitz and Stegun (1972), it is known that

$$\lim_{\kappa \to 0} D_n(\kappa) = 1 \quad \text{and} \quad D_n(\kappa) \approx \frac{n}{\kappa^n} \Gamma(n+1)\zeta(n+1)$$

for large positive κ, where $\Gamma(\cdot)$ and $\zeta(\cdot)$ are gamma and Riemann zeta functions, respectively. So in the special case of $n = 1$, we readily have

$$\lim_{\kappa \to 0} D_1(\kappa) = 1 \quad \text{and} \quad D_1(\kappa) \approx \frac{\pi^2}{6\kappa} \tag{10.15}$$

for large positive κ, since $\Gamma(2) = 1$ and $\zeta(2) = \sum_{l=1}^{\infty} \frac{1}{l^2} = \frac{\pi^2}{6}$.

Thus, from (10.12) and the properties of Debye function stated in (10.15), we see that Kendall's tau measure will be 0 when $\kappa \to 0$, and will be approximately

$$1 - \frac{4}{\kappa}\left(1 - \frac{\pi^2}{6\kappa}\right), \quad \text{or simply,} \quad 1 - \frac{4}{\kappa}, \tag{10.16}$$

when κ is large.

An exact series expression can also be provided for Kendall's tau measure in this case, and is given by

$$\tau = 1 - \frac{4}{\kappa} + \frac{2\pi^2}{3\kappa^2} - \frac{4}{\kappa}\sum_{l=0}^{\infty}\frac{\exp\{-(l+1)\kappa\}}{l+1} - \frac{4}{\kappa^2}\sum_{l=0}^{\infty}\frac{\exp\{-(l+1)\kappa\}}{(l+1)^2}. \quad (10.17)$$

The detailed derivation of the expression in (10.17) is presented in Appendix F.

Here again, we assume that $T_{i,g}$ has its marginal probability distribution as $F_{i,g}(t)$, and that the dependence parameter is related to stress level x_i in a linear form as

$$\kappa_i = d_0 + d_1 x_i, \quad i = 1, 2, \dots, I. \quad (10.18)$$

Then, the joint cumulative distribution function of $T_{i,1}$ and $T_{i,2}$, under the considered copula, can be expressed from (10.10) as

$$\begin{aligned} &C_{\kappa_i}(F_{i,1}(t_1), F_{i,2}(t_2)) \\ &= P(T_{i,1} \le t_1, T_{i,2} \le t_2) \\ &= -\frac{1}{\kappa_i}\ln\left\{1 + \frac{[\exp(-\kappa_i F_{i,1}(t_1)) - 1][\exp(-\kappa_i F_{i,2}(t_2)) - 1]}{(\exp(-\kappa_i) - 1)}\right\}. \end{aligned} \quad (10.19)$$

When κ_i approaches ∞, $T_{i,1}$ and $T_{i,2}$ become highly positively associated. Also, if $T_{i,1}$ is less than $T_{i,2}$ in the stochastic order, i.e. $R_{i,1}(t) \le R_{i,2}(t)$ for all $t > 0$, we will get in the limit (as association becomes highly positively stronger)

$$R_i(t) = R_{i,1}(t) - F_{i,2}(t) + \lim_{\kappa_i \to \infty} C_{\kappa_i}(F_{i,1}(t), F_{i,2}(t)) = R_{i,1}(t). \quad (10.20)$$

On the other hand, when κ_i approaches $-\infty$, $T_{i,1}$ and $T_{i,2}$ become highly negatively associated. Furthermore, for $0 \le F_{i,1}(t) \le 1$ and $0 \le F_{i,2}(t) \le 1$, we obtain in the limit (as association becomes highly negatively stronger)

$$\begin{aligned} R_i(t) &= R_{i,1}(t) - F_{i,2}(t) + \lim_{\kappa_i \to -\infty} C_{\kappa_i}(F_{i,1}(t), F_{i,2}(t)) \\ &= R_{i,1}(t) - F_{i,2}(t) + \max(0, F_{i,1}(t) + F_{i,2}(t) - 1) \\ &= \max(1 - F_{i,1}(t) - F_{i,2}(t), 0). \end{aligned} \quad (10.21)$$

These are the upper and lower Fréchet–Hoeffding copula bounds; see Kotz et al. (2000).

It is evident from (10.21) that the reliability of the device is $\max(1 - F_1(t) - F_2(t), 0)$ when the times for the two failure modes are highly negatively associated, while the reliability of the device is $R_1(t)R_2(t) = 1 - F_1(t) - F_2(t) + F_1(t)F_2(t)$ when the times for the two failure modes are independent. We also note that devices with two independent failure modes would have longer lifetimes than those with highly negatively associated failure modes. This makes sense in that a failure mode with a longer lifetime would provide information that the time for the other failure mode would likely be short when the two times of failure are highly negatively associated.

10.4 Estimation of Dependence

Let $C_{\theta_C}(i,j) = C_{\kappa_i(\theta_C)}(F_{i,1}(\tau_j), F_{i,2}(\tau_j))$. Let us use $p_{i,j,12}, p_{i,j,1}, p_{i,j,2}$ and $p_{i,j,0}$ to denote the probabilities of both failure modes, of failure mode 1 only and of failure mode 2 only, and the reliability, respectively. We then readily have

$$\begin{aligned} p_{i,j,12} &= C_{\theta_C}(i,j), \\ p_{i,j,1} &= F_{i,1}(\tau_j) - C_{\theta_C}(i,j), \\ p_{i,j,2} &= F_{i,2}(\tau_j) - C_{\theta_C}(i,j), \\ p_{i,j,0} &= 1 - F_{i,1}(\tau_j) - F_{i,2}(\tau_j) + C_{\theta_C}(i,j). \end{aligned}$$

With the above notation, the log-likelihood function, based on the observed data $\mathbf{z}$, can be expressed as

$$\begin{aligned} \ell(\theta_C) = \sum_{i=1}^{I}\sum_{j=1}^{J} \{ & n_{i,j,12}\ln(p_{i,j,12}) + n_{i,j,1}\ln(p_{i,j,1}) + n_{i,j,2}\ln(p_{i,j,2}) \\ & + n_{i,j,0}\ln(p_{i,j,0})\} . \end{aligned} \tag{10.22}$$

We now present a two-step procedure for parameter estimation, in which the marginal distributions are first estimated by the proportions of failures, i.e. $\hat{F}_{i,g}(\tau_j) = N_{i,j,g}/K_{i,j}$, which are then plugged into the log-likelihood to obtain a composite log-likelihood from (10.22) as follows:

$$\begin{aligned} \ell(\theta_C) = \sum_{i=1}^{I}\sum_{j=1}^{J} \Big\{ & n_{i,j,12}\ln(C_{\theta_C}(i,j)) + n_{i,j,1}\ln\left(\hat{F}_{i,1}(\tau_j) - C_{\theta_C}(i,j)\right) \\ & + n_{i,j,2}\ln\left(\hat{F}_{i,2}(\tau_j) - C_{\theta_C}(i,j)\right) \\ & + n_{i,j,0}\ln\left(1 - \hat{F}_{i,1}(\tau_j) - \hat{F}_{i,2}(\tau_j) + C_{\theta_C}(i,j)\right)\Big\}, \end{aligned} \tag{10.23}$$

where $\theta_C = (c_0, c_1)$ for the Gumbel–Hougaard copula, while $\theta_C = (d_0, d_1)$ for the Frank copula. An estimate of model parameter θ_C can then be obtained by solving a system of composite log-likelihood equations for θ_C derived from (10.23). Then, the solution to the system of equations is the quasimaximum likelihood estimate (QMLE) of model parameter. Because the estimate cannot be expressed in explicit form, optimization algorithms, such as optim() in R and fminsearch() in MATLAB, can be used to find the estimate. It needs to be observed that $\hat{F}_{i,s}(\tau_j) = C_{\theta_C}(i,j)$ when $\hat{F}_{i,g}(\tau_j) = 1$ for $s \neq g$, which results in the term $\ln(\hat{F}_{i,s}(\tau_j) - C_{\theta_C}(i,j))$ in the composite log-likelihood function in (10.23) being undefined. To overcome this problem, we suggest replacing 1 by $K_{i,j}(0.99^{(1/K_{i,j})})$ for $\hat{F}_{i,g}(\tau_j)$ as an adjustment. Similarly, for the Gumbel–Hougaard copula, $C_{\theta_C}(i,j)$ becomes undefined in (10.23) when $\hat{F}_{i,g}(\tau_j) = 0$, and here again we suggest replacing 0 by $K_{i,j}(1 - 0.99^{(1/K_{i,j})})$ for $\hat{F}_{i,g}(\tau_j)$ as an adjustment. In short, suitable adjustments are needed when either (i) $N_{i,j,g} = 0$ or (ii) $N_{i,j,g} = K_{i,j}$, for $g = 1, 2, i = 1, 2, \ldots, I, j = 1, 2, \ldots, J$.

10.5 Simulation Studies

Along the lines of Ling et al. (2020), we conduct here an extensive Monte Carlo simulation study to examine the performance of the proposed estimation method for the Gumbel–Hougaard and Frank copulas with weak ($\tau = 0.17$ or 0.18) and moderate ($\tau = 0.38$ or 0.39) dependence, with Weibull and gamma lifetime distributions. We evaluate in terms of bias and root mean square error (RMSE) of the QMLE of model parameter $\boldsymbol{\theta}_C$ of the dependence parameter as well as the dependence parameter κ_0 under normal operating condition x_0, based on various sample sizes $K = 50$ (small), $K = 100$ (moderate), and $K = 200$ (large).

The experiment scenario we considered for this study consists of three elevated stress levels ($x_1 = 30, x_2 = 40, x_3 = 50$) and four inspection times ($\tau_j \in \{5, 10, 15, 20\}$). At stress level x_i and inspection time τ_j, K devices were selected and tested. The dependence parameter κ_0, under normal operating condition $x_0 = 25$, was determined based on the observed data. Under weak dependence, the parameters of interest under (i) Gumbel–Hougaard copula, (ii) Frank copula with positive dependence, and (iii) Frank copula with negative dependence were taken as $(c_0, c_1, \kappa_0) = (-2, 0.02, 1.22)$, $(d_0, d_1, \kappa_0) = (1, 0.02, 1.50)$, and $(d_0, d_1, \kappa_0) = (-1, -0.02, -1.50)$, respectively. Under moderate dependence, the parameters of interest were correspondingly taken as $(c_0, c_1, \kappa_0) = (-1, 0.02, 1.61)$, $(d_0, d_1, \kappa_0) = (3, 0.02, 3.50)$, and $(d_0, d_1, \kappa_0) = (-3, -0.02, -3.50)$. In addition, the lifetime distributions corresponding to the two failure modes were taken to be Weibull and gamma, and so they served as the marginal distributions in the joint distributions given by (10.6) and (10.10). Specifically, we assumed the marginal distribution of T_g to be Weibull with shape parameter $\eta_{i,g} = \exp(r_{0,g} + r_1 x_i)$ and a common scale parameter $\beta_i = \exp(s_0 + s_1 x_i)$, where $(r_{0,1}, r_{0,2}, r_1, s_0, s_1) = (2.0, 2.1, -0.03, 3.5, -0.02)$. Similarly, we assumed the marginal distribution of T_g to be gamma with shape parameter $\alpha_{i,g} = \exp(a_{0,g} + a_1 x_i)$ and a common scale parameter $\beta_i = \exp(b_0 + b_1 x_i)$, where $(a_{0,1}, a_{0,2}, a_1, b_0, b_1) = (3.6, 3.8, -0.06, -0.3, 0.04)$. All the results obtained from this simulation study, based on 10 000 Monte Carlo simulations, are summarized in Tables 10.2 and 10.3.

The results in Tables 10.2 and 10.3 show that the bias and RMSE of the parameters of interest, under Weibull and gamma lifetime distributions, are similar for all considered cases, regardless of the copula model used, thus suggesting that the estimation method for the dependence parameter is quite robust. As a result, the effect of misspecification of marginal distribution on the estimation of the dependence parameter is negligible. We also observe that the bias and RMSE decrease when the sample size increases for all considered cases. Surprisingly, for Frank copula, we observe from Tables 10.2 and 10.3 that the estimation has a larger RMSE for moderate dependence than the corresponding RMSE for weak dependence; on

Table 10.2 Bias and root mean square error (RMSE) of the QMLEs of parameters $(c_0, c_1, d_0, d_1, \kappa_0)$ in the case under Gumbel–Hougaard (GH) copula ($\kappa_0 = 1.22$ and $\tau = 0.18$ at $x_0 = 25$) and Frank ($\kappa_0 = \pm 1.50$ and $\tau = \pm 0.17$ at $x_0 = 25$) copula with weak dependence, with Weibull and gamma lifetime distributions, for various sample sizes.

GH Parameter	Bias $K = 50$	Bias $K = 100$	Bias $K = 200$	RMSE $K = 50$	RMSE $K = 100$	RMSE $K = 200$
$c_0 = -2.0$[a]	0.3703	0.1876	0.0876	3.4322	1.3453	0.6065
$c_0 = -2.0$[b]	0.3585	0.2369	0.1267	5.0185	1.8029	0.6237
$c_1 = 0.02$[a]	−0.0194	−0.0068	−0.0023	0.0100	0.0019	0.0004
$c_1 = 0.02$[b]	−0.0218	−0.0082	−0.0030	0.0107	0.0026	0.0004
$\kappa_0 = 1.22$[a]	0.1087	0.0508	0.0224	0.0487	0.0175	0.0070
$\kappa_0 = 1.22$[b]	0.1103	0.0570	0.0277	0.0511	0.0198	0.0079

Frank Parameter	Bias $K = 50$	Bias $K = 100$	Bias $K = 200$	RMSE $K = 50$	RMSE $K = 100$	RMSE $K = 200$
$d_0 = 1.00$[a]	0.6692	0.3234	0.1585	7.0923	3.4313	1.6927
$d_0 = 1.00$[b]	0.7736	0.4060	0.1899	7.3092	3.5751	1.6996
$d_1 = 0.02$[a]	−0.0092	−0.0051	−0.0029	0.0040	0.0020	0.0010
$d_1 = 0.02$[b]	−0.0127	−0.0075	−0.0037	0.0041	0.0020	0.0010
$\kappa_0 = 1.50$[a]	0.4393	0.1970	0.0850	1.3723	0.6297	0.3029
$\kappa_0 = 1.50$[b]	0.4552	0.2196	0.0975	1.4054	0.6704	0.3103

Frank Parameter	Bias $K = 50$	Bias $K = 100$	Bias $K = 200$	RMSE $K = 50$	RMSE $K = 100$	RMSE $K = 200$
$d_0 = -1.00$[a]	0.8822	0.4809	0.2269	5.4363	3.1071	1.7217
$d_0 = -1.00$[b]	1.1353	0.6506	0.3175	6.1211	3.4248	1.7537
$d_1 = -0.02$[a]	−0.0048	−0.0047	−0.0030	0.0028	0.0017	0.0010
$d_1 = -0.02$[b]	−0.0126	−0.0099	−0.0056	0.0030	0.0019	0.0010
$\kappa_0 = -1.50$[a]	0.7616	0.3631	0.1527	1.4042	0.6360	0.3162
$\kappa_0 = -1.50$[b]	0.8205	0.4035	0.1776	1.5349	0.7005	0.3260

a) Weibull
b) Gamma

Table 10.3 Bias and root mean square error (RMSE) of the QMLEs of parameters $(c_0, c_1, d_0, d_1, \kappa_0)$ in the case under Gumbel–Hougaard copula ($\kappa_0 = 1.61$ and $\tau = 0.38$ at $x_0 = 25$) and Frank copula ($\kappa_0 = \pm 3.50$ and $\tau = \pm 0.39$ at $x_0 = 25$) with moderate dependence, with Weibull and gamma lifetime distributions, for various sample sizes.

GH	Bias			RMSE		
Parameter	$K = 50$	$K = 100$	$K = 200$	$K = 50$	$K = 100$	$K = 200$
$c_0 = -1.00$[a]	0.2418	0.1159	0.0690	0.9919	0.4915	0.2473
$c_0 = -1.00$[b]	0.1813	0.1350	0.0844	0.9558	0.5286	0.2682
$c_1 = 0.02$[a]	−0.0048	−0.0021	−0.0013	0.0006	0.0003	0.0001
$c_1 = 0.02$[b]	−0.0044	−0.0025	−0.0016	0.0005	0.0003	0.0001
$\kappa_0 = 1.61$[a]	0.1371	0.0686	0.0366	0.1131	0.0446	0.0199
$\kappa_0 = 1.61$[b]	0.1000	0.0764	0.0428	0.0947	0.0505	0.0230

Frank	Bias			RMSE		
Parameter	$K = 50$	$K = 100$	$K = 200$	$K = 50$	$K = 100$	$K = 200$
$d_0 = 3.00$[a]	0.6591	0.3825	0.1775	11.0991	5.3085	2.5008
$d_0 = 3.00$[b]	0.5688	0.4591	0.2102	10.3493	5.6078	2.7005
$d_1 = 0.02$[a]	−0.0087	−0.0061	−0.0028	0.0064	0.0031	0.0015
$d_1 = 0.02$[b]	−0.0088	−0.0079	−0.0037	0.0060	0.0032	0.0016
$\kappa_0 = 3.50$[a]	0.4413	0.2300	0.1076	2.0637	0.9690	0.4460
$\kappa_0 = 3.50$[b]	0.3487	0.2614	0.1173	1.9075	1.0563	0.4948

Frank	Bias			RMSE		
Parameter	$K = 50$	$K = 100$	$K = 200$	$K = 50$	$K = 100$	$K = 200$
$d_0 = -3.00$[a]	1.5600	0.8252	0.4264	6.6751	3.8749	2.4113
$d_0 = -3.00$[b]	1.7785	1.0137	0.5957	7.7842	4.4303	2.6393
$d_1 = -0.02$[a]	−0.0035	−0.0028	−0.0027	0.0026	0.0019	0.0013
$d_1 = -0.02$[b]	−0.0113	−0.0091	−0.0079	0.0029	0.0021	0.0014
$\kappa_0 = -3.50$[a]	1.4737	0.7558	0.3585	2.9139	1.1321	0.5193
$\kappa_0 = -3.50$[b]	1.4956	0.7870	0.3975	3.0421	1.2109	0.5531

a) Weibull
b) Gamma

the other hand, for Gumbel–Hougaard copula, the case of moderate dependence always results in a smaller RMSE than the corresponding RMSE for the case of weak dependence. This may be due to the fact that Gumbel–Hougaard copula allows only positive dependence, while Frank copula allows both positive and negative dependence.

10.6 Case Study with R Codes

In this section, we consider serial sacrifice data presented in Table 1.7 for studying onset time and the rate of development of radiation-induced disease for the purpose of illustrating all the results developed in the preceding sections. We are primarily concerned with the presence or absence of two disease categories – (i) thymic lymphoma and/or glomerulosclerosis and (ii) all other diseases – for an irradiated group ($x = 1$) of 343 female mice given γ-radiation and control group ($x = 0$) of 361 radiation-free female mice. The R codes for defining the data comprising two disease categories and for the estimation of model parameters are presented in Tables 10.4 and 10.5, respectively.

The QMLEs and 95% percentile bootstrap confidence intervals for the parameters under Gumbel–Hougaard and Frank copulas are presented in Table 10.6. These results show that the confidence intervals for a_1 under models M1 and M3 include 0, indicating that the presence of γ-radiation does not change the dependence parameters. The Akaike information criterion (AIC) and the Bayesian information criterion (BIC) also enable us to conclude that the presence of γ-radiation does not have any effect on the dependence parameter. We, therefore, assume the dependence parameter to be constant over stress levels, and thus obtain reduced models M2 and M4. Both AIC and BIC show that model M4 fits the data best among all considered models, and this model, having positive association, shows that the onset times of the two disease categories are associated positively, meaning that one disease category induces the other disease category.

Table 10.4 R codes for defining the data with two correlated disease categories in Table 1.7.

```
X<-rep(c(0,1),each=7)
n0<-c(58,40,18,8,1,1,0,54,36,13,0,0,0,0)
n1<-c(13,23,41,25,21,11,9,12,24,35,13,3,0,0)
n2<-c(0,1,1,1,1,0,1,1,3,1,2,1,1,1)
n12<-c(1,1,3,6,16,21,39,0,5,17,28,35,30,28)
n<-matrix(c(n0,n1,n2,n12),ncol=4)
```

Table 10.5 R codes for the estimation of model parameters under Gumbel–Hougaard and Frank copulas.

```
source("Copula.R")
#Gumbel-Hougaard copula (M1)
MG2<-OS.Copula(n,X,M='Gumbel',Cov=T)
#Gumbel-Hougaard copula (M2)
MG<-OS.Copula(n,X,M='Gumbel',Cov=F) #Frank copula (M3)
MF2<-OS.Copula(n,X,M='Frank',Cov=T)
#Frank copula (M4)
MF<-OS.Copula(n,X,M='Frank',Cov=F)
#Independence model (Ind)
MInd<-OS.Copula(n,X,M='Ind',Cov=F)
N<-n[X==0,]+n[X==1,]
F1<-apply(N[,c(2,4)],1,sum)/apply(N,1,sum)
F2<-apply(N[,c(3,4)],1,sum)/apply(N,1,sum)
a.hat<-1+exp(MG$a.MLE)
GH<-exp(-((-log(F1))^a.hat+(-log(F2))^a.hat)^(1/a.hat))
a.hat<-MF$a.MLE
FR<-log(1+(exp(-a.hat*F1)-1)*(exp(-a.hat*F2)-1)/(exp(-a.hat)-1))/a.hat
```

Table 10.6 QMLEs and 95% percentile bootstrap confidence intervals of parameters under Gumbel–Hougaard (GH) and Frank copulas, along with AIC and BIC values.

Model	Copula	QMLE	95% CI	AIC	BIC
M1[a]	GH	$\hat{c}_0 = -2.135$	$[-42.468, -0.456]$	1059.964	1069.077
		$\hat{c}_1 = 0.048$	$[-13.461, 13.904]$		
M2[b]	GH with $c_1 = 0$	$\hat{\kappa} = 1.121$	$[1, 1.384]$	1058.178	1062.735
M3[c]	Frank	$\hat{d}_0 = 1.342$	$[-0.949, 5.134]$	1058.108	1067.222
		$\hat{d}_1 = 0.425$	$[-0.374, 3.888]$		
M4[d]	Frank with $d_1 = 0$	$\hat{\kappa} = 1.574$	$[0.020, 3.438]$	1056.248	1060.804

a) MG2$theta.MLE.
b) MG$a.MLE.
c) MF2$theta.MLE.
d) MF$a.MLE.

Table 10.7 Estimates of the probabilities of absence of disease category (a), disease category (b), and both disease categories under different models.

τ	Absence of (a) $R_1(\tau)$	Absence of (b) $R_2(\tau)$	Absence of (a) and (b)		
			Ind[a)]	M2[b)]	M4[c)]
100	0.8129	0.9856	0.8013	0.8126	0.8032
200	0.6015	0.9248	0.5563	0.6008	0.5694
300	0.2558	0.8295	0.2122	0.2558	0.2310
400	0.1325	0.5542	0.0735	0.1325	0.0945
500	0.0385	0.3205	0.0123	0.0385	0.0189
600	0.0312	0.1875	0.0059	0.0312	0.0099
700	0.0256	0.1154	0.0030	0.0256	0.0053

a) $(1 - F1) * (1 - F2)$.
b) $1 - F1 - F2 + \mathrm{GH}$.
c) $1 - F1 - F2 + \mathrm{FR}$.

Under the independence assumption on the two onset times, 10 000 samples of $(n_{i,j,0}, n_{i,j,1}, n_{i,j,2}, n_{i,j,12})$ were generated, each of which was used to compute the QMLE of κ under both Gumbel–Hougaard and Frank copulas. We then found that 705 samples had the QMLE greater than the observed estimate of $\hat{\kappa} = 1.1209$ for Gumbel–Hougaard copula, while only 241 samples had the QMLE greater than the observed estimate of $\hat{\kappa} = 1.5738$ for Frank copula. These seem to support the hypothesis that the onset times of the two disease categories are likely not independent. It is worth noting that the system reliability at a mission time, at different stress levels, can also be estimated under the copula models in a similar manner as done above with the use of two copula models, and bear in mind that the system reliability in the present context corresponds to the case of onset times for both disease being larger than the inspection time.

Table 10.7 presents the estimates of the probabilities of absence of disease category (a), disease category (b), and both disease categories under independence, Gumbel–Hougaard copula (M2) and Frank copula (M4) models. Under the reduced models M2 and M4, with constant dependence parameters for both groups, the numbers of mice $(n_{j,0}, n_{j,1}, n_{j,2}, n_{j,12})$ are combined to estimate the probabilities of the absence of disease category (a) $(R_1(\tau_j))$, disease category (b) $(R_2(\tau_j))$ and both disease categories $(R(\tau_j))$. We observe that the probabilities for both M2 and M4 models are greater than those for the independence model due to the positive association between the onset times of the two disease categories. A final observation that needs to be made here is that, though the onset times of the two disease categories are positively associated, the presence of thymic lymphoma and/or glomerulosclerosis is considerably more likely than all other diseases at all inspection times considered.

11

Conclusions and Future Directions

11.1 Brief Overview

In the preceding chapters, we have discussed statistical inference for one-shot devices for different lifetime distributions from the frequentist, Bayesian, and robust inferential perspectives. Optimal designs of constant-stress and step-stress accelerated life-tests for one-shot devices have also been discussed. Finally, inferential methods for one-shot devices with multiple components, when the components function independently as well as in a dependent manner, have also been developed. In this chapter, we first provide some concluding remarks and then point out some further problems of interest in this direction.

11.2 Concluding Remarks

11.2.1 Large Sample Sizes for Flexible Models

The lifetime data observed from one-shot devices are all either left- or right-censored. For this reason, data analysis of one-shot device lifetimes poses a real challenge in developing efficient inferential methods. When a flexible lifetime distribution with more model parameters is considered for modeling the data, the likelihood function becomes quite complicated, with its maximization becoming a complex numerical problem. If the number of devices tested is not large enough, there may be numerical instability in the required maximization process. In such a case, the numerical iterative process for determining the global maxima (namely, the maximum likelihood estimates of model parameters) may not converge. If some prior information is available from past experiments in such situations, Bayesian approach and the associated machinery would become quite

Accelerated Life Testing of One-shot Devices: Data Collection and Analysis, First Edition.
Narayanaswamy Balakrishnan, Man Ho Ling, and Hon Yiu So.

Companion Website: www.wiley.com/go/Balakrishnan/Accelerated_Life_Testing

useful. Having said that, it will always be desirable to have a large sample size in order to estimate the model parameters as well as the associated reliability characteristics in a precise manner, and also to avoid potential computational problems during the estimation process.

11.2.2 Accurate Estimation

The selection of suitable link functions to relate stress levels to specific parameters of a lifetime distribution is quite critical for accurate estimation of reliability and mean lifetime under normal operating conditions. Many commonly used reliability models, such as Arrhenius, power law, inverse power law, and Eyring models, are all of the log-linear form with appropriate transformation on the stress variables. Depending on the nature and purpose of the reliability experiment on one-shot devices and the stress variables used in the experiment, reliability engineers may wish to consider the use of specific link functions (probably based on some physical and/or thermodynamic principles) in the analysis. It is important to mention that this will always be possible, with all the methodologies described in the preceding chapters suitably modified in this case. However, the assumption of a wrong lifetime distribution (that is model mis-specification) may result in serious bias in the estimates. For avoiding this, we would recommend the use of specification tests to validate the lifetime distribution assumed for analyzing the data. Furthermore, contaminated data are often encountered in reliability practice. In this case, the use of robust estimators would help mitigate the effect of data contamination and result in robust estimates of parameters when there are outliers present in the data, but at a price of a small loss in efficiency when there are in fact no outliers in the data.

11.2.3 Good Designs Before Data Analysis

Many one-shot devices in practice, such as vehicle airbags and missiles, are expensive and so practitioners may only have limited number of devices for testing purposes. For this reason, design of accelerated life-tests becomes very important for securing data efficiently in a relatively short period of time. In the context of one-shot devices, this becomes especially important as all lifetimes of devices are censored; therefore, determining a good design and implementing it in the life-test would result in reducing a substantial number of test devices for the accelerated life-test, which in turn would decrease the cost of the experiment while achieving the specified standard error.

11.3 Future Directions

11.3.1 Weibull Lifetime Distribution with Threshold Parameter

In reliability studies, it may be often reasonable to assume that there is a nonzero origin below which no event can occur; in such situations, the Weibull distribution with an unknown location (threshold) parameter will be more suitable for the analysis. However, the usual regularity conditions for the maximum likelihood estimation of parameters will not be satisfied in this case as the support of the distribution depends on the unknown threshold parameter; in fact, the maximum likelihood estimators may not even exist in certain circumstances. In this case, it will naturally be of great interest to develop alternative estimation methods for the parameters of Weibull lifetime distribution with threshold based on one-shot device testing data. One good possibility may be to utilize the method of modified likelihood estimation developed recently by Nagatsuka et al. (2013) and Nagatsuka and Balakrishnan (2013, 2015, 2016) and adopt it suitably to the case of one-shot device testing data.

11.3.2 Frailty Models

To model dependence among multiple failure modes of one-shot devices, we used the theory of copulas. Though there is a plethora of choices of copulas for modeling purpose, we focused our discussion on the specific copulas – independence, Frank, and Gumbel–Hougaard copulas. Another prominent way to model dependence is by means of frailty models. Frailty models will not only enable us in capturing interrelationships between multiple failure modes but will also facilitate in describing the influence of latent factors, and providing easily interpretable results. Yet, another problem of interest will be regarding the selection of "best copula" to model a given one-shot device testing data.

11.3.3 Optimal Design of SSALTs with Multiple Stress Levels

In Chapter 8, optimal design of simple step-stress accelerated life-test (SSALTs) with two stress levels have been discussed in detail; but, optimal design of SSALTs with more than two stress levels still remains an open problem. This situation will naturally involve more variables in the optimization problem for the determination of inspection times and allocation of devices while minimizing the asymptotic variance of a reliability parameter of interest.

11.3.4 Comparison of CSALTs and SSALTs

In Chapters 7 and 8, optimal design of constant-stress accelerated life-test (CSALTs) and SSALTs for one-shot devices have been described in detail. SSALTs are usually more efficient and require fewer devices for testing than CSALTs; however, CSALTs are easy to implement in a life-test for data collection. It will, therefore, be of great interest to develop suitable methods for evaluating the amount of additional information that can be gained from SSALTs and at what additional cost.

Appendix A

Derivation of $H_i(a, b)$

We have

$$
\begin{aligned}
H_i(a,b) &= \frac{1}{\Gamma(a)} \int_0^b (\ln(x))^i x^{a-1} \exp(-x) dx \\
&= \frac{1}{\Gamma(a)} \frac{\partial^i}{\partial a^i} \int_0^b x^{a-1} \exp(-x) dx \\
&= \frac{1}{\Gamma(a)} \frac{\partial^i}{\partial a^i} \int_0^b x^{a-1} \sum_{n=0}^{\infty} \frac{(-x)^n}{n!} dx \\
&= \frac{1}{\Gamma(a)} \frac{\partial^i}{\partial a^i} \int_0^b \sum_{n=0}^{\infty} \frac{(-1)^n x^{n+a-1}}{n!} dx \\
&= \frac{1}{\Gamma(a)} \sum_{n=0}^{\infty} (-1)^n \frac{\partial^i}{\partial a^i} \left(\frac{b^{n+a}}{n!(n+a)} \right).
\end{aligned} \tag{A.1}
$$

Therefore,

$$
\begin{aligned}
H_1(a,b) &= \frac{1}{\Gamma(a)} \sum_{n=0}^{\infty} (-1)^n \frac{\partial}{\partial a} \left(\frac{b^{n+a}}{n!(n+a)} \right) \\
&= \frac{1}{\Gamma(a)} \sum_{n=0}^{\infty} \frac{(-1)^n b^{n+a}}{n!(n+a)} \left(\ln(b) - \frac{1}{n+a} \right) \\
&= \frac{\ln(b)}{\Gamma(a)} \sum_{n=0}^{\infty} \frac{(-1)^n b^{n+a}}{n!(n+a)} - \frac{1}{\Gamma(a)} \sum_{n=0}^{\infty} \frac{(-1)^n b^{n+a}}{n!(n+a)^2} \\
&= \ln(b)\gamma(a,b) - \frac{b^a {}_2F_2(a,a;a+1,a+1;-b)}{a^2\Gamma(a)},
\end{aligned} \tag{A.2}
$$

and

$$
H_2(a,b) = \frac{1}{\Gamma(a)} \sum_{n=0}^{\infty} (-1)^n \frac{\partial^2}{\partial a^2} \left(\frac{b^{n+a}}{n!(n+a)} \right)
$$

Accelerated Life Testing of One-shot Devices: Data Collection and Analysis, First Edition.
Narayanaswamy Balakrishnan, Man Ho Ling, and Hon Yiu So.

Companion Website: www.wiley.com/go/Balakrishnan/Accelerated_Life_Testing

$$= \frac{1}{\Gamma(a)} \sum_{n=0}^{\infty} \frac{(-1)^n b^{n+a}}{n!} \left(\frac{(\ln(b))^2}{n+1} - \frac{2\ln(b)}{(n+a)^2} + \frac{2}{(n+a)^3} \right)$$

$$= (\ln(b))^2 \gamma(a, b) - \frac{2\ln(b) b^{a_2 F_2}(a, a; a+1, a+1; -b)}{a^{2\Gamma(a)}}$$

$$+ \frac{2b^a {}_3F_3(a, a, a; a+1, a+1, a+1; -b)}{a^3\Gamma(a)}; \tag{A.3}$$

in the above expressions, ${}_AF_B(a_1, a_2, \ldots, a_A; b-1, b_2, \ldots, b_B; z)$ denotes the generalized hypergeometric function defined by (see Slater 2008)

$${}_AF_B(a_1, a_2, \ldots, a_A; b-1, b_2, \ldots, b_B; z)$$

$$= 1 + \sum_{k=0}^{\infty} \frac{\prod_{j=0}^{k(a_1+j)(a_2+j)\cdots(a_A+j)}}{\prod_{j=0}^{k(b_1+j)(b_2+j)\cdots(b_B+j)}} \frac{z^{k+1}}{(k+1)!}. \tag{A.4}$$

Thus, for example the function ${}_2F_2(a, a; a+1; a+1; z)$ appearing in (A.2) is simply given by

$${}_2F_2(a, a; a+1; a+1; z) = 1 + \sum_{k=0}^{\infty} \left(\frac{a}{a+1+k} \right)^2 \frac{z^{k+1}}{(k+1)!}. \tag{A.5}$$

Appendix B

Observed Information Matrix

To obtain the observed information matrix under expectation-maximization (EM) framework, the missing information principle requires the complete information matrix and the missing information matrix given by

$$I_{\text{complete}} = -E\left[\frac{\partial^2(\ell_c(\boldsymbol{\theta}))}{\partial\boldsymbol{\theta}\partial\boldsymbol{\theta}'}\right] \quad \text{and} \quad I_{\text{missing}} = -E\left[\frac{\partial^2(\ln(f(t_{ik}|\mathbf{z},\boldsymbol{\theta})))}{\partial\boldsymbol{\theta}\partial\boldsymbol{\theta}'}\right].$$

Using these matrices, we will then obtain the observed information matrix as

$$I_{\text{obs}}(\boldsymbol{\theta}) = I_{\text{complete}} - I_{\text{missing}}.$$

Suppose the lifetime distribution has a probability density function (pdf) $f(t;\boldsymbol{\theta})$ and a cumulative distribution function (cdf) $F(t;\boldsymbol{\theta})$. Given data with a sequence of inspection times $0 < \tau_1 < \tau_2 < \cdots < \tau_{I-1}$, the corresponding numbers of failures within each of these time intervals, $(n_1, n_2, \ldots, n_I)$, and K observed failure times, $t_k, k = 1, 2, \ldots, K$, the log-likelihood function for observed data is then given by

$$\ell(\boldsymbol{\theta}) = n_1 \ln(F(\tau_1;\boldsymbol{\theta})) + \sum_{i=2}^{I-1} n_i \ln(F(\tau_i;\boldsymbol{\theta}) - F(\tau_{i-1};\boldsymbol{\theta}))$$

$$+ n_I \ln(1 - F(\tau_{I-1};\boldsymbol{\theta})) + \sum_{k=1}^{K} \ln(f(t_k;\boldsymbol{\theta})) + \text{constant},$$

where $n_1, n_2, \ldots, n_I$ and $t_1, t_2, \ldots, t_K$ are all random variables.

On the other hand, the log-likelihood function for complete data is given by

$$\ell_{\text{complete}} = \sum_{i=1}^{I}\sum_{j=1}^{n_i} \ln(f(t_j;\boldsymbol{\theta})) + \sum_{k=1}^{K} \ln(f(t_k;\boldsymbol{\theta})) + \text{constant},$$

and the log-likelihood function, from the conditional distribution for the missing data, is given by

$$\ell_{\text{missing}} = \sum_{j=1}^{n_1}(\ln(f(t_j;\boldsymbol{\theta})) - \ln(F(\tau_1;\boldsymbol{\theta})))$$

Accelerated Life Testing of One-shot Devices: Data Collection and Analysis, First Edition.
Narayanaswamy Balakrishnan, Man Ho Ling, and Hon Yiu So.

Companion Website: www.wiley.com/go/Balakrishnan/Accelerated_Life_Testing

$$
\begin{aligned}
&+\sum_{i=2}^{I-1}\sum_{j=1}^{n_i}(\ln(f(t_j;\boldsymbol{\theta}))-\ln(F(\tau_i;\boldsymbol{\theta})-F(\tau_{i-1};\boldsymbol{\theta})))\\
&+\sum_{j=1}^{n_I}(\ln(f(t_j;\boldsymbol{\theta}))-\ln(1-F(\tau_{I-1};\boldsymbol{\theta})))+\text{constant}\\
=&\sum_{i=1}^{I}\sum_{j=1}^{n_i}\ln(f(t_j;\boldsymbol{\theta}))-\sum_{j=1}^{n_1}\ln(F(\tau_1;\boldsymbol{\theta}))\\
&-\sum_{i=2}^{I-1}\sum_{j=1}^{n_i}\ln(F(\tau_i;\boldsymbol{\theta})-F(\tau_{i-1};\boldsymbol{\theta}))-\sum_{j=1}^{n_I}\ln(1-F(\tau_{I-1};\boldsymbol{\theta}))\\
&+\text{constant}.
\end{aligned}
$$

Note that t_j in ℓ_{complete} and ℓ_{missing} are also random variables, but the terms of $\ln(f(t_j;\boldsymbol{\theta}))$ get canceled out by the missing information principle. Hence, the observed log-likelihood function obtained from the missing information principle is

$$
\begin{aligned}
\ell_{\text{obs}}(\boldsymbol{\theta})=&\,n_1\ln(F(\tau_1;\boldsymbol{\theta}))+\sum_{i=2}^{I-1}n_i\ln(F(\tau_i;\boldsymbol{\theta})-F(\tau_{i-1};\boldsymbol{\theta}))\\
&+n_I\ln(1-F(\tau_{I-1};\boldsymbol{\theta}))+\sum_{k=1}^{K}\ln(f(t_k;\boldsymbol{\theta}))+\text{constant},
\end{aligned}
$$

where $t_1, t_2, \ldots, t_K$ are random variables.

For the case when all lifetimes are censored, the terms of $\sum_{k=1}^{K}\ln(f(t_k;\boldsymbol{\theta}))$ in both $\ell(\boldsymbol{\theta})$ and ℓ_{complete} disappear, and thus, $\ell_{\text{obs}}(\boldsymbol{\theta})=\ell(\boldsymbol{\theta})$. Moreover, it is easy to show that the observed information matrix from the missing information principle is identical to the Fisher information matrix based on the observed likelihood function, see Balakrishnan and Ling (2013).

For the case of one-shot device testing data wherein all failure times are censored, the observed information matrix, by the missing information principle, is then given by

$$
\ell(\boldsymbol{\theta})=\sum_{i=1}^{I}n_i\ln(F(\tau_i;\boldsymbol{\theta}))+(K_i-n_i)\ln(1-F(\tau_i;\boldsymbol{\theta}))+\text{constant}.
$$

So, the second-order derivatives with respect to $\boldsymbol{\theta}$ are

$$
\begin{aligned}
\frac{\partial^2(\ell(\boldsymbol{\theta}))}{\partial\boldsymbol{\theta}\partial\boldsymbol{\theta}'}=&\sum_{i=1}^{I}\left(\frac{\partial^2F(\tau_i;\boldsymbol{\theta})}{\partial\boldsymbol{\theta}\partial\boldsymbol{\theta}'}\right)\left(\frac{n_i}{F(\tau_i;\boldsymbol{\theta})}-\frac{K_i-n_i}{1-F(\tau_i;\boldsymbol{\theta})}\right)\\
&-\sum_{i=1}^{I}\left(\frac{\partial F(\tau_i;\boldsymbol{\theta})}{\partial\boldsymbol{\theta}}\right)
\end{aligned}
$$

$$\times \left(\frac{\partial F(\tau_i;\boldsymbol{\theta})}{\partial \boldsymbol{\theta}'} \right) \left(\frac{n_i}{(F(\tau_i;\boldsymbol{\theta}))^2} + \frac{K_i - n_i}{(1 - F(\tau_i;\boldsymbol{\theta}))^2} \right).$$

Thus, we finally obtain the observed information matrix, which is the expectation of the negative of the second-order derivatives of the observed log-likelihood function with respect to the model parameters, as

$$\begin{aligned} I_{\text{obs}}(\boldsymbol{\theta}) &= -E\left[\frac{\partial^2(\ell(\boldsymbol{\theta}))}{\partial \boldsymbol{\theta} \partial \boldsymbol{\theta}'} \right] \\ &= -\sum_{i=1}^{I} \left(\frac{\partial^2 F(\tau_i;\boldsymbol{\theta})}{\partial \boldsymbol{\theta} \partial \boldsymbol{\theta}'} \right) \left(\frac{E[n_i]}{F(\tau_i;\boldsymbol{\theta})} - \frac{K_i - E[n_i]}{1 - F(\tau_i;\boldsymbol{\theta})} \right) \\ &\quad + \sum_{i=1}^{I} \left(\frac{\partial F(\tau_i;\boldsymbol{\theta})}{\partial \boldsymbol{\theta}} \right) \left(\frac{\partial F(\tau_i;\boldsymbol{\theta})}{\partial \boldsymbol{\theta}'} \right) \left(\frac{E[n_i]}{(F(\tau_i;\boldsymbol{\theta}))^2} + \frac{K_i - E[n_i]}{(1 - F(\tau_i;\boldsymbol{\theta}))^2} \right) \\ &= \sum_{i=1}^{I} K_i \left(\frac{\partial F(\tau_i;\boldsymbol{\theta})}{\partial \boldsymbol{\theta}} \right) \left(\frac{\partial F(\tau_i;\boldsymbol{\theta})}{\partial \boldsymbol{\theta}'} \right) \left(\frac{1}{F(\tau_i;\boldsymbol{\theta})} + \frac{1}{1 - F(\tau_i;\boldsymbol{\theta})} \right). \end{aligned}$$

Appendix C

Non-Identifiable Parameters for SSALTs Under Weibull Distribution

Given one-shot device testing data under simple step-stress accelerated life-tests (SSALTs), only the numbers of failures are observed, and the likelihood function would then involve just $R(\tau_1)$ and $R(\tau_2)$. Under the cumulative exposure (CE) model, let

$$H_1 = \ln(R(\tau_1)) = \left(\frac{\tau_1}{\beta_1}\right)^{\eta} \quad \text{and} \quad H_2 = \ln(R(\tau_2)) = \left(\frac{\tau_1}{\beta_1} + \frac{\tau_2 - \tau_1}{\beta_2}\right)^{\eta}.$$

It then follows that

$$\beta_1 = \frac{\tau_1}{H_1^{\frac{1}{\eta}}} \quad \text{and} \quad \beta_2 = \frac{\tau_2 - \tau_1}{H_2^{\frac{1}{\eta}} - H_1^{\frac{1}{\eta}}}.$$

As it is assumed that $\beta_i = \exp(s_0 + s_1 x_i)$, we subsequently obtain

$$s_1 = \frac{\ln\left(H_1^{\frac{1}{\eta}}\right) - \ln\left(H_2^{\frac{1}{\eta}} - H_1^{\frac{1}{\eta}}\right) + \ln(\tau_2 - \tau_1) - \ln(\tau_1)}{x_2 - x_1},$$

$$s_0 = \left(\frac{x_1}{x_2 - x_1}\right)\left\{\ln\left(H_2^{\frac{1}{\eta}} - H_1^{\frac{1}{\eta}}\right) - \ln(\tau_2 - \tau_1)\right\} - \ln\left(H_1^{\frac{1}{\eta}}\right) + \ln(\tau_1).$$

From the above expressions, it is observed that both s_0 and s_1 are functions of η, implying that there does not exist a unique $\boldsymbol{\theta}_W = (s_0, s_1, \eta)$ maximizing the likelihood function. Hence, the model parameter $\boldsymbol{\theta}_W$ is not identifiable for one-shot device testing data under simple SSALTs.

Accelerated Life Testing of One-shot Devices: Data Collection and Analysis, First Edition.
Narayanaswamy Balakrishnan, Man Ho Ling, and Hon Yiu So.

Companion Website: www.wiley.com/go/Balakrishnan/Accelerated_Life_Testing

Appendix D

Optimal Design Under Weibull Distributions with Fixed w_1

For optimal design with fixed a_1, the sample size relies on the estimated reliability and the limit on standard deviation in this setup, regardless of the stress levels. This is because the asymptotic standard deviation gets minimized when either $\pi_2 = 1$ or $\pi_2 = 0$, implying that, in (8.26), either

$$\frac{\tau_1}{\beta_1} = \frac{t}{\beta_0} \quad \text{or} \quad \frac{\tau_1}{\beta_1} + \frac{\tau_2 - \tau_1}{\beta_2} = \frac{t}{\beta_0}.$$

Consequently,

$$C_2 = \sqrt{R(t)(1 - R(t))}$$

and

$$K = \frac{R(t)(1 - R(t))}{SD^2}.$$

Accelerated Life Testing of One-shot Devices: Data Collection and Analysis, First Edition.
Narayanaswamy Balakrishnan, Man Ho Ling, and Hon Yiu So.

Companion Website: www.wiley.com/go/Balakrishnan/Accelerated_Life_Testing

Appendix E

Conditional Expectations for Competing Risks Model Under Exponential Distribution

First, for g = 1, 2, we have

$$\begin{aligned} E[T_{i,g,k} | \{T_{i,g,k} > \tau_i\}] &= \frac{\int_{\tau_i}^{\infty} t\lambda_{i,g} \exp(-\lambda_{i,g} t)dt}{\exp(-\lambda_{i,g}\tau_i)} \\ &= \exp(\lambda_{i,g}\tau_i) \left\{ [-t\exp(-\lambda_{i,g}t)]_{\tau_i}^{\infty} + \int_{\tau_i}^{\infty} \exp(-\lambda_{i,g}t)dt \right\} \\ &= \exp(\lambda_{i,g}\tau_i) \left\{ \tau_i \exp(-\lambda_{i,g}\tau_i) + \frac{\exp(-\lambda_{i,g}\tau_i)}{\lambda_{i,g}} \right\} \\ &= \tau_i + \frac{1}{\lambda_{i,g}}. \end{aligned}$$

The following formula is useful in the subsequent derivations:

$$\begin{aligned} \int_0^A t\lambda \exp(-\lambda t)dt &= [-t\exp(-\lambda t)]_0^A + \int_0^A \exp(-\lambda t)dt \\ &= -A\exp(-\lambda A) + \frac{1-\exp(-\lambda A)}{\lambda}. \end{aligned}$$

For the probability for the case $\Delta = 1$, we obtain:

$$\begin{aligned} P(T_{i,1,k} < \min(\tau_i, T_{i,2,k})) &= \int_0^{\infty} \int_0^{\min(\tau_i,t_2)} f_1(t_1)f_2(t_2)dt_1dt_2 \\ &= \int_0^{\tau_i} \int_0^{t_2} \lambda_{i,1} \exp(-\lambda_{i,1}t_1)\lambda_{i,2}\exp(-\lambda_{i,2}t_2)dt_1dt_2 \\ &\quad + \int_0^{\tau_i} \lambda_{i,1}\exp(-\lambda_{i,1}t_1)dt_1 \int_{\tau_i}^{\infty} \lambda_{i,2}\exp(-\lambda_{i,2}t_2)dt_2 \\ &= \int_0^{\tau_i} (1-\exp(\lambda_{i,1}t_2))\lambda_{i,2}\exp(-\lambda_{i,2}t_2)dt_2 \\ &\quad + \int_0^{\tau_i} \lambda_{i,1}\exp(-\lambda_{i,1}t_1)dt_1 \int_{\tau_i}^{\infty} \lambda_{i,2}\exp(-\lambda_{i,2}t_2)dt_2 \end{aligned}$$

Accelerated Life Testing of One-shot Devices: Data Collection and Analysis, First Edition.
Narayanaswamy Balakrishnan, Man Ho Ling, and Hon Yiu So.

Companion Website: www.wiley.com/go/Balakrishnan/Accelerated_Life_Testing

$$
\begin{aligned}
&= 1 - \exp(-\lambda_{i,2}\tau_i) - \left(\frac{\lambda_{i,2}}{\lambda_{i,1} + \lambda_{i,2}}\right) \\
&\quad \times (1 - \exp(-(\lambda_{i,1} + \lambda_{i,2})\tau_i)) \\
&\quad + \exp(-\lambda_{i,2}\tau_i) - \exp(-(\lambda_{i,1} + \lambda_{i,2})\tau_i) \\
&= \left(\frac{\lambda_{i,1}}{\lambda_{i,1} + \lambda_{i,2}}\right)\{1 - \exp(-(\lambda_{i,1} + \lambda_{i,2})\tau_i)\} \\
&= P(T_{i,1,k} < T_{i,2,k})P(\min(T_{i,1,k}, T_{i,2,k}) < \tau_i).
\end{aligned}
$$

Similarly, we find

$$
\begin{aligned}
P(T_{i,2,k} < \min(\tau_i, T_{i,1,k})) &= \left(\frac{\lambda_{i,2}}{\lambda_{i,1} + \lambda_{i,2}}\right)\{1 - \exp(-(\lambda_{i,1} + \lambda_{i,2})\tau_i)\} \\
&= P(T_{i,2,k} < T_{i,1,k})P(\min(T_{i,1,k}, T_{i,2,k}) < \tau_i).
\end{aligned}
$$

We then find

$$
\begin{aligned}
&E[T_{i,1,k} | \{T_{i,1,k} < \min(\tau_i, T_{i,2,k})\}] \\
&= \frac{1}{P(T_{i,1,k} < \min(\tau_i, T_{i,2,k}))} \int_0^{\infty} \int_0^{\min(\tau_i, t_2)} t_1 f_1(t_1) f_2(t_2) dt_1 dt_2 \\
&= \frac{1}{P(T_{i,1,k} < \min(\tau_i, T_{i,2,k}))} \int_0^{\tau_i} \int_0^{t_2} \lambda_{i,1} t_1 \exp(-\lambda_{i,1} t_1) \lambda_{i,2} \exp(-\lambda_{i,2} t_2) dt_1 dt_2 \\
&\quad + \frac{1}{P(T_{i,1,k} < \min(\tau_i, T_{i,2,k}))} \int_0^{\tau_i} \lambda_{i,1} t_1 \exp(-\lambda_{i,1} t_1) dt_1 \\
&\times \int_{\tau_i}^{\infty} \lambda_{i,2} \exp(-\lambda_{i,2} t_2) dt_2 \\
&= \frac{1}{P(T_{i,1,k} < \min(\tau_i, T_{i,2,k}))} \int_0^{\tau_i} -\lambda_{i,2} t_2 \exp(-(\lambda_{i,1} + \lambda_{i,2}) t_2) \\
&\quad + \frac{\lambda_{i,2}}{\lambda_{i,1}} \exp(-\lambda_{i,2} t_2) dt_2 \\
&\quad - \frac{1}{P(T_{i,1,k} < \min(\tau_i, T_{i,2,k}))} \int_0^{\tau_i} \frac{\lambda_{i,2}}{\lambda_{i,1}} \exp(-(\lambda_{i,1} + \lambda_{i,2}) t_2) dt_2 \\
&\quad - \frac{1}{P(T_{i,1,k} < \min(\tau_i, T_{i,2,k}))} \tau_i \exp(-(\lambda_{i,1} + \lambda_{i,2})\tau_i) \\
&\quad + \frac{\exp(-\lambda_{i,2}\tau_i) - \exp(-(\lambda_{i,1} + \lambda_{i,2})\tau_i)}{\lambda_{i,1} P(T_{i,1,k} < \min(\tau_i, T_{i,2,k}))} \\
&= \frac{1}{P(T_{i,1,k} < \min(\tau_i, T_{i,2,k}))} \left[\frac{\lambda_{i,2}}{\lambda_{i,1} + \lambda_{i,2}}\right. \\
&\quad \times \left\{\tau_i \exp(-(\lambda_{i,1} + \lambda_{i,2})\tau_i) - \frac{1 - \exp(-(\lambda_{i,1} + \lambda_{i,2})\tau_i)}{\lambda_{i,1} + \lambda_{i,2}}\right\}
\end{aligned}
$$

$$+\frac{1}{\lambda_{i,1}}(1-\exp(-\lambda_{i,2}\tau_i)) - \frac{\lambda_{i,1}}{\lambda_{i,1}(\lambda_{i,1}+\lambda_{i,2})}(1-\exp(-(\lambda_{i,1}+\lambda_{i,2})\tau_i))$$
$$\left. -\tau_i \exp(-(\lambda_{i,1}+\lambda_{i,2})\tau_i) + \frac{1}{\lambda_{i,1}}\exp(-\lambda_{i,2}\tau_i) - \frac{1}{\lambda_{i,1}}\exp(-(\lambda_{i,1}+\lambda_{i,2})\tau_i)\right]$$
$$= \frac{1}{\lambda_{i,1}+\lambda_{i,2}} - \frac{\tau_i \exp(-(\lambda_{i,1}+\lambda_{i,2})\tau_i)}{1-\exp(-(\lambda_{i,1}+\lambda_{i,2})\tau_i)}.$$

Similarly, we find

$$E[T_{i,2,k} | \{T_{i,2,k} < \min(\tau_i, T_{i,1,k})\}] = \frac{1}{\lambda_{i,1}+\lambda_{i,2}} - \frac{\tau_i \exp(-(\lambda_{i,1}+\lambda_{i,2})\tau_i)}{1-\exp(-(\lambda_{i,1}+\lambda_{i,2})\tau_i)}$$

and

$$E[T_{i,2,k} | \{T_{i,1,k} < \min(\tau_i, T_{i,2,k})\}]$$
$$= \frac{1}{P(T_{i,1,k} < \min(\tau_i, T_{i,2,k}))} \int_0^\infty \int_0^{\min(\tau_i, t_2)} t_2 f_1(t_1) f_2(t_2) dt_1 dt_2$$
$$= \frac{1}{P(T_{i,1,k} < \min(\tau_i, T_{i,2,k}))} \int_0^{\tau_i} \int_0^{t_2} \lambda_{i1} \exp(-\lambda_{i,1}t_1) dt_1 \lambda_{i,2} t_2 \exp(-\lambda_{i,2}t_2) dt_2$$
$$+ \frac{1}{P(T_{i,1,k} < \min(\tau_i, T_{i,2,k}))} \int_0^{\tau_i} \lambda_{i,1} \exp(-\lambda_{i,1}t_1) dt_1$$
$$\times \int_{\tau_i}^\infty t_2 \lambda_{i,2} \exp(-\lambda_{i,2}t_2) dt_2$$
$$= \frac{1}{P(T_{i,1,k} < \min(\tau_i, T_{i,2,k}))} \int_0^{\tau_i} \{1-\exp(-\lambda_{i,1}t_2)\} t_2 \lambda_{i,2} \exp(-\lambda_{i,2}t_2) dt_2$$
$$+ \left(\tau_i + \frac{1}{\lambda_{i,2}}\right) \frac{\exp(-\lambda_{i,2}\tau_i)\{1-\exp(-\lambda_{i,1}\tau_i)\}}{P(T_{i,1,k} < \min(\tau_i, T_{i,2,k}))}$$
$$= \frac{1}{P(T_{i,1,k} < \min(\tau_i, T_{i,2,k}))} \left[\frac{1-\exp(-\lambda_{i,2}\tau_i)}{\lambda_{i,2}} - \tau_i \exp(-\lambda_{i,2}\tau_i)\right.$$
$$\left. - \frac{\lambda_{i,2}}{\lambda_{i,1}+\lambda_{i,2}} \left\{ \frac{1-\exp(-(\lambda_{i,1}+\lambda_{i,2})\tau_i)}{\lambda_{i,1}+\lambda_{i,2}} - \tau_i \exp(-(\lambda_{i,1}+\lambda_{i,2})\tau_i) \right\}\right]$$
$$+ \left(\tau_i + \frac{1}{\lambda_{i,2}}\right) \frac{\exp(-\lambda_{i,2}\tau_i)(1-\exp(-\lambda_{i,1}\tau_i))}{P(T_{i,1,k} < \min(\tau_i, T_{i,2,k}))}$$
$$= \frac{1}{\lambda_{i,2}} + \frac{1}{\lambda_{i,1}+\lambda_{i,2}} - \frac{\tau_i \exp(-(\lambda_{i,1}+\lambda_{i,2})\tau_i)}{1-\exp(-(\lambda_{i,1}+\lambda_{i,2})\tau_i)}.$$

In an analogous manner, we also find

$$E[T_{i,1,k} | \{T_{i,2,k} < \min(\tau_i, T_{i,1,k})\}] = \frac{1}{\lambda_{i,1}} + \frac{1}{\lambda_{i,1}+\lambda_{i,2}}$$
$$- \frac{\tau_i \exp(-(\lambda_{i,1}+\lambda_{i,2})\tau_i)}{1-\exp(-(\lambda_{i,1}+\lambda_{i,2})\tau_i)}.$$

Next, for the case with masked cause of failure, we obtain

$$
\begin{aligned}
E[T_1|\Delta = -1] &= E[T_1|\{\min(T_1, T_2) \le \tau\}] \\
&= \frac{1}{P(\min(T_1, T_2) \le \tau)} \\
&\quad \times \int\int_{\min(T_1,T_2)<\tau} t_1 \exp(-\lambda_1 t_1) t_2 \exp(-\lambda_2 t_2) dt_1 dt_2 \\
&= \frac{1}{P(\min(T_1, T_2) \le \tau)} \\
&\quad \times \left\{ \int_0^\infty \int_0^\infty t_1 \exp(-\lambda_1 t_1) t_2 \exp(-\lambda_2 t_2) dt_1 dt_2 \right. \\
&\quad \left. - \int\int_{\min(T_1,T_2)\ge\tau} t_1 \exp(-\lambda_1 t_1) t_2 \exp(-\lambda_2 t_2) dt_1 dt_2 \right\} \\
&= \frac{1}{P(\min(T_1, T_2) \le \tau)} \\
&\quad \times \left\{ \frac{1}{\lambda_1} - \exp(-\lambda_2 \tau) \int_\tau^\infty t_1 \exp(-\lambda_1 t_1) dt_1 \right\} \\
&= \frac{1}{1 - \exp(-(\lambda_1 + \lambda_2)\tau)} \\
&\quad \times \left\{ \frac{1}{\lambda_1} - \left(\tau + \frac{1}{\lambda_1} \right) \exp(-(\lambda_1 + \lambda_2)\tau) \right\} \\
&= \frac{1}{\lambda_1} - \frac{\tau \exp(-(\lambda_1 + \lambda_2)\tau)}{1 - \exp(-(\lambda_1 + \lambda_2)\tau)}.
\end{aligned}
$$

Similarly, we find

$$
E[T_2|\Delta = -1] = E[T_2|\{\min(T_1, T_2) \le \tau\}] = \frac{1}{\lambda_2} - \frac{\tau \exp(-(\lambda_1 + \lambda_2)\tau)}{1 - \exp(-(\lambda_1 + \lambda_2)\tau)}.
$$

Appendix F

Kendall's Tau for Frank Copula

First, let us consider the Debye function

$$\begin{aligned}
D_1(\kappa) &= \frac{1}{\kappa}\int_0^{\kappa} \frac{t}{\exp(t)-1}dt \\
&= \frac{1}{\kappa}\int_0^{\kappa} \frac{t\exp(-t)}{1-\exp(-t)}dt \\
&= \frac{1}{\kappa}\sum_{l=0}^{\infty}\int_0^{\kappa} t\exp\{-(l+1)t\}dt \\
&= -\frac{1}{\kappa}\sum_{l=0}^{\infty}\frac{1}{l+1}\int_0^{\kappa} td(\exp\{-(l+1)t\}) \\
&= -\frac{1}{\kappa}\sum_{l=0}^{\infty}\left\{\frac{1}{l+1}[t\exp\{-(l+1)t\}]_0^{\kappa} - \int_0^{\kappa}\exp\{-(l+1)t\}dt\right\} \\
&= -\frac{1}{\kappa}\sum_{l=0}^{\infty}\frac{1}{l+1}\left\{\kappa\exp\{-(l+1)\kappa\} + \left[\frac{\exp\{-(l+1)t\}}{l+1}\right]_0^{\kappa}\right\} \\
&= -\frac{1}{\kappa}\sum_{l=0}^{\infty}\frac{1}{l+1}\left\{\kappa\exp\{-(l+1)\kappa\} + \frac{\exp\{-(l+1)\kappa\}-1}{l+1}\right\} \\
&= -\sum_{l=0}^{\infty}\frac{\exp\{-(l+1)\kappa\}-1}{l+1} - \frac{1}{\kappa}\sum_{l=0}^{\infty}\frac{\exp\{-(l+1)\kappa\}-1}{(l+1)^2} \\
&\quad + \frac{1}{\kappa}\sum_{l=0}^{\infty}\frac{1}{(l+1)^2} \\
&= \frac{\pi^2}{6\kappa} - \sum_{l=0}^{\infty}\frac{\exp\{-(l+1)\kappa\}}{l+1} - \frac{1}{\kappa}\sum_{l=0}^{\infty}\frac{\exp\{-(l+1)\kappa\}}{(l+1)^2}. \qquad \text{(F.1)}
\end{aligned}$$

From the expression in (F.1), we also readily observe the approximation $D_1(\kappa) \approx \frac{\pi^2}{6\kappa}$ for large κ, mentioned earlier in Chapter 10.

Accelerated Life Testing of One-shot Devices: Data Collection and Analysis, First Edition.
Narayanaswamy Balakrishnan, Man Ho Ling, and Hon Yiu So.

Companion Website: www.wiley.com/go/Balakrishnan/Accelerated_Life_Testing

Upon substituting the expression of $D_1(\kappa)$ in (F.1) into the formula for Kendall's tau measure, we obtain an exact series expression for it as

$$\tau = 1 - \frac{4}{\kappa} + \frac{2\pi^2}{3\kappa^2} - \frac{4}{\kappa}\sum_{l=0}^{\infty}\frac{\exp(-(l+1)\kappa)}{l+1} - \frac{4}{\kappa^2}\sum_{l=0}^{\infty}\frac{\exp(-(l+1)\kappa)}{(l+1)^2}. \tag{F.2}$$

Bibliography

M. Abramowitz and I. A. Stegun. *Handbook of Mathematical Functions, with Formulas, Graphs and Mathematical Tables*. US Government Printing Office, Washington, DC, 1972.

H. Akaike. On entropy maximization principle. In P. R. Krishnaiah, editor, *Applications of Statistics*, pages 27–41. North-Holland, Amsterdam, 1977.

H. Akaike. On entropy maximization principle likelihood of a model and information criteria. *Journal of Econometrics*, 16:3–14, 1981.

A. A. Alhadeed and S. S. Yang. Optimal simple step-stress plan for cumulative exposure model using log-normal distribution. *IEEE Transactions on Reliability*, 54:64–68, 2005.

F. J. Anscombe. The transformation of Poisson, binomial and negative-binomial data. *Biometrika*, 35:246–254, 1948.

R. Antoine, H. Doss, and M. Hollander. On identifiability in the autopsy model of reliability theory. *Journal of Applied Probability*, 30:913–930, 1993.

M. Aslam and C. H. Jun. A group acceptance sampling plan for truncated life test having Weibull distribution. *Journal of Applied Statistics*, 36:1021–1027, 2009.

D. S. Bai, M. S. Kim, and S. H. Lee. Optimum simple step-stress accelerated life tests with censoring. *IEEE Transactions on Reliability*, 38:528–532, 1989.

L. J. Bain and M. Engelhardt. Reliability test plans for one-shot devices based on repeated samples. *Journal of Quality Technology*, 23:304–311, 1991.

N. Balakrishnan. Progressive censoring methodology: An appraisal (with discussions). *Test*, 16:211–296, 2007.

N. Balakrishnan. A synthesis of exact inferential results for exponential step-stress models and associated optimal accelerated life tests. *Metrika*, 69: 351–396, 2008.

N. Balakrishnan and A. P. Basu. (*Eds.*) *The Exponential Distribution: Theory, Methods and Applications*. Taylor and Francis, Philadelphia, 1995.

Accelerated Life Testing of One-shot Devices: Data Collection and Analysis, First Edition.
Narayanaswamy Balakrishnan, Man Ho Ling, and Hon Yiu So.

Companion Website: www.wiley.com/go/Balakrishnan/Accelerated_Life_Testing

N. Balakrishnan and E. Chimitova. Goodness-of-fit tests for one-shot device accelerated life testing data. *Communications in Statistics - Simulation and Computation*, 46:3723–3734, 2017.

N. Balakrishnan and A. C. Cohen. *Order Statistics and Inference: Estimation Methods*. Academic Press, Boston, 1991.

N. Balakrishnan and E. Cramer. *The Art of Progressive Censoring: Applications to Reliability and Quality*. Birkhäuser, Boston, 2014.

N. Balakrishnan and C. D. Lai. *Continuous Bivariate Distributions*, Second edition. Springer, New York, 2010.

N. Balakrishnan and M. H. Ling. EM algorithm for one-shot device testing under the exponential distribution. *Computational Statistics & Data Analysis*, 56: 502–509, 2012a.

N. Balakrishnan and M. H. Ling. Multiple-stress model for one-shot device testing data under exponential distribution. *IEEE Transactions on Reliability*, 61:809–821, 2012b.

N. Balakrishnan and M. H. Ling. Expectation maximization algorithm for one shot device accelerated life testing with Weibull lifetimes, and variable parameters over stress. *IEEE Transactions on Reliability*, 62:537–551, 2013.

N. Balakrishnan and M. H. Ling. Gamma lifetimes and one-shot device testing analysis. *Reliability Engineering & System Safety*, 126:54–64, 2014a.

N. Balakrishnan and M. H. Ling. Best constant-stress accelerated life-test plans with multiple stress factors for one-shot device testing under a Weibull distribution. *IEEE Transactions on Reliability*, 63:944–952, 2014b.

N. Balakrishnan, D. Kundu, H. K. T. Ng, and N. Kannan. Point and interval estimation for a simple step-stress model with Type-II censoring. *Journal of Quality Technology*, 39:35–47, 2007.

N. Balakrishnan, H. Y. So, and M. H. Ling. EM algorithm for one-shot device testing with competing risks under exponential distribution. *Reliability Engineering & System Safety*, 137:129–140, 2015.

N. Balakrishnan, H. Y. So, and M. H. Ling. A Bayesian approach for one-shot device testing with exponential lifetimes under competing risks. *IEEE Transactions on Reliability*, 65:469–485, 2016a.

N. Balakrishnan, H. Y. So, and M. H. Ling. EM algorithm for one-shot device testing with competing risks under Weibull distribution. *IEEE Transactions on Reliability*, 65:973–991, 2016b.

N. Balakrishnan, E. Castilla, N. Martin, and L. Pardo. Robust estimators and test statistics for one-shot device testing under the exponential distribution. *IEEE Transactions on Information Theory*, 65:3080–3096, 2019a.

N. Balakrishnan, E. Castilla, N. Martin, and L. Pardo. Robust estimators for one-shot device testing data under gamma lifetime model with an application to a tumor toxicological data. *Metrika*, 82:991–1019, 2019b.

N. Balakrishnan, E. Castilla, N. Martin, and L. Pardo. Robust inference for one-shot device testing data under Weibull lifetime model. *IEEE Transactions on Reliability*, 69:937–953, 2020a.

N. Balakrishnan, E. Castilla, N. Martin, and L. Pardo. Robust inference for one-shot device testing data under exponential lifetime model with multiple stresses. *Quality and Reliability Engineering International*, 36:1916–1930, 2020b.

N. Balakrishnan, E. Castilla, N. Martin, and L. Pardo. Statistical inference for one-shot devices based on density power divergences: An overview. In B.C. Arnold, N. Balakrishnan, and C. Coelho, editors, *Contributions to Statistical Distribution Theory and Inference – Festschrift in Honor of C. R. Rao on the Occasion of His 100th Birthday*. Springer, New York, 2020c.

G. Barmalzan, S. M. Ayat, N. Balakrishnan, and R. Roozegar. Stochastic comparisons of series and parallel systems with dependent heterogeneous extended exponential components under Archimedean copula. *Journal of Computational and Applied Mathematics*, 380:112965, 2020.

S. Basak, A. Basu, and M. C. Jones. On the 'optimal' density power divergence tuning parameter. *Journal of Applied Statistics*, 2020.

A. Basu, I. R. Harris, N. L. Hjort, and M. C. Jones. Robust and efficient estimation by minimising a density power divergence. *Biometrika*, 85:549–559, 1998.

A. Basu, H. Shioya, and C. Park. *Statistical Inference: The Minimum Distance Approach*. Chapman and Hall/CRC Press, New York, 2011.

A. Basu, A. Mandal, N. Martin, and L. Pardo. Generalized Wald-type tests based on minimum density power divergence estimators. *Statistics*, 50:1–26, 2016.

A. Basu, A. Ghosh, N. Martin, and L. Pardo. Robust Wald-type tests for non-homogeneous observations based on the minimum density power divergence estimators. *Metrika*, 81:493–522, 2018.

S. Basu. Masked failure data: competing risks. In F. Ruggeri, R. S. Kenett, and F. W. Faltin, editors, *Encyclopedia of Statistics in Quality and Reliability*. John Wiley & Sons, Chichester, England, 2008.

B. Berlin, J. Brodsky, and P. Clifford. Testing disease dependence in survival experiments with serial sacrifice. *Journal of the American Statistical Association*, 74:5–14, 1979.

G. K. Bhattacharyya and Z. Soejoeti. A tampered failure rate model for step-stress accelerated life test. *Communications in Statistics - Theory and Methods*, 18:1627–1643, 1989.

G. Casella and R. L. Berger. *Statistical Inference, Second edition*. Brooks/Cole, Boston, 2002.

Y. Cheng and E. A. Elsayed. Reliability modeling and prediction of systems with mixture of units. *IEEE Transactions on Reliability*, 65:914–928, 2016.

Y. Cheng and E. A. Elsayed. Reliability modeling of mixtures of one-shot units under thermal cyclic stresses. *Reliability Engineering & System Safety*, 167: 58–66, 2017.

Y. Cheng and E. A. Elsayed. Reliability modeling and optimization of operational use of one-shot units. *Reliability Engineering & System Safety*, 176:27–36, 2018.

U. Cherubini, E. Luciano, and W. Vecchiato. *Copula Methods in Finance*. John Wiley & Sons, Hoboken, New Jersey, 2004.

F. Y. Chiyoshi. Modeling dependence with copulas: a useful tool for field development decision process. *Journal of Petroleum Science and Engineering*, 44:83–91, 2018.

A. C. Cohen. *Truncated and Censored Samples: Theory and Applications*. Marcel Dekker, New York, 1991.

D. R. Cox. Regression models and life tables. *Journal of the Royal Statistical Society: Series B*, 34:187–220, 1972.

D. R. Cox and D. Oakes. *Analysis of Survival Data*. Chapman and Hall, London, England, 1984.

R. V. Craiu and T. Duchesne. Inference based on the EM algorithm for the competing risks model with masked causes of failure. *Biometrika*, 91:543–558, 2004.

E. Cramer and U. Kamps. Sequential *k*-out-*n* systems. In N. Balakrishnan and C. R. Rao, editors, *Handbook of Statistics Vol. 20, Advances in Reliability*, pages 301–372. North-Holland, Amsterdam, 2001.

M. H. DeGroot and P. K. Goel. Bayesian estimation and optimal designs in partially accelerated life testing. *Naval Research Logistics Quarterly*, 26: 223–235, 1979.

L. A. Escobar and W. Q. Meeker. Planning accelerated life tests with two or more experimental factors. *Technometrics*, 37:411–427, 1995.

T. H. Fan, N. Balakrishnan, and C. C. Chang. The Bayesian approach for highly reliable electro-explosive devices using one-shot device testing. *Journal of Statistical Computation and Simulation*, 79:1143–1154, 2009.

L. Fang, N. Balakrishnan, and Q. Jin. Optimal grouping of heterogeneous components in series-parallel and parallel-series systems under Archimedean copula dependence. *Journal of Computational and Applied Mathematics*, 377: 112916, 2020.

W. Feller. *An Introduction to Probability Theory and Its Applications - Vol. I*. John Wiley & Sons, New York, 1971.

M. Fréchet. Les tableaux de corrlation et les programmes liné aires. *Revue de l'Institut international de statistique*, pages 23–40, 1957.

C. Genest and J. MacKay. The joy of copulas: bivariate distributions with uniform marginals. *The American Statistician*, 40:280–283, 1986.

A. Ghosh and A. Basu. Robust estimation for independent non-homogeneous observations using density power divergence with applications to linear regression. *Electronic Journal of Statistics*, 7:2420–2456, 2013.

E. Gouno. An inference method for temperature step-stress accelerated life testing. *Quality and Reliability Engineering International*, 17:11–18, 2001.

E. Gouno. Optimum step stress for temperature accelerated life testing. *Quality and Reliability Engineering International*, 23:915–924, 2007.

H. Guo, S. Honecker, A. Mettas, and D. Ogden. Reliability estimation for one-shot systems with zero component test failures. In *Proceedings-Annual Reliability and Maintainability Symposium, RAMS*, pages 1–7, 2010.

F. R. Hampel, E. M. Ronchetti, P. J. Rousseeuw, and W. A. Stahel. *Robust Statistics: The Approach based on Influence Functions*. John Wiley & Sons, 1986.

W. K. Hastings. Monte Carlo sampling methods using Markov chains and their applications. *Biometrika*, 57:97–109, 1970.

H. P. Hong, W. Zhou, S. Zhang, and W. Ye. Optimal condition-based maintenance decisions for systems with dependent stochastic degradation of components. *Reliability Engineering & System Safety*, 121:276–288, 2014.

M. H. Hoyle. Transformations: an introduction and a bibliography. *International Statistical Review*, 41:203–223, 1973.

D. Huard, G. Évin, and A. C. Favre. Bayesian copula selection. *Computational Statistics & Data Analysis*, 51:809–822, 2006.

C. P. Hwang and H. Y. Ke. A reliability analysis technique for quantal-response data. *Reliability Engineering & System Safety*, 41:239–243, 1993.

X. Jia, L. Wang, and C. Wei. Reliability research of dependent failure systems using copula. *Communications in Statistics - Simulation and Computation*, 43:1838–1851, 2014.

H. Joe. *Dependence Modeling with Copulas*. CRC Press, Boca Raton, Florida, 2014.

N. L. Johnson, S. Kotz, and N. Balakrishnan. *Continuous Univariate Distributions - Vol 1, Second edition*. John Wiley & Sons, New York, 1994.

N. L. Johnson, S. Kotz, and N. Balakrishnan. *Continuous Univariate Distributions - Vol 2*, Second edition. John Wiley & Sons, New York, 1995.

M. Kendall. A new measure of rank correlation. *Biometrika*, 30:81–89, 1938.

D. K. Kim and J. M. Taylor. Transform-both-sides approach for overdispersed binomial data when *N* is unobserved. *Journal of the American Statistical Association*, 89:833–845, 1994.

R. L. Kodell and C. J. Nelson. An illness-death model for the study of the carcinogenic process using survival/sacrifice data. *Biometrics*, 36:267–277, 1980.

S. Kotz, N. Balakrishnan, and N. L. Johnson. *Continuous Multivariate Distributions - Vol.1, Second edition*. John Wiley & Sons, New York, 2000.

S. Kullback and R. A. Leibler. On information and sufficiency. *The Annals of Mathematical Statistics*, 22:79–86, 1951.

J. H. Lau, G. Harkins, D. Rice, J. Kral, and B. Wells. Experimental and statistical analyses of surface-mount technology PLCC solder-joint reliability. *IEEE Transactions on Reliability*, 37:524–530, 1988.

C. T. Lin, Y. Y. Hsu, S. Y. Lee, and N. Balakrishnan. Inference on constant stress accelerated life tests for log-location-scale lifetime distributions with Type-I hybrid censoring. *Journal of Statistical Computation and Simulation*, 89:720–749, 2019.

J. C. Lindsey and L. M. Ryan. A three state multiplicative model for rodent tumorigenicity experiments. *Journal of the Royal Statistical Society: Series C*, 42:283–300, 1993.

M. H. Ling. Optimal design of simple step-stress accelerated life test for one-shot devices under exponential distributions. *Probability in the Engineering and Informational Sciences*, 33:121–135, 2019.

M. H. Ling and N. Balakrishnan. Model mis-specification analyses of Weibull and gamma models for one-shot device testing data. *IEEE Transactions on Reliability*, 66:641–650, 2017.

M. H. Ling and X. W. Hu. Optimal design of simple step-stress accelerated life test for one-shot devices under Weibull distributions. *Reliability Engineering & System Safety*, 193:106630, 2020.

M. H. Ling, H. Y. So, and N. Balakrishnan. Likelihood inference under proportional hazards model for one-shot device testing. *IEEE Transactions on Reliability*, 65:446–458, 2016.

M. H. Ling, P. S. Chan, H. K. T. Ng, and N. Balakrishnan. Copula models for one-shot device testing data with correlated failure modes. *Communications in Statistics - Theory and Methods*, 2020.

T. A. Louis. Finding the observed information matrix when using the EM algorithm. *Journal of the Royal Statistical Society: Series B*, 44:226–233, 1982.

G. J. McLachlan and T. Krishnan. *The EM Algorithm and Extensions*, Second edition. John Wiley & Sons, Hoboken, New Jersey, 2008.

W. Q. Meeker and L. A. Escobar. *Statistical Methods for Reliability Data*. John Wiley & Sons, Hoboken, New Jersey, 2014.

W. Q. Meeker and G. J. Hahn. *How to Plan an Accelerated Life Test: Some Practical Guidelines*. American Society for Quality Control, Milwaukee, 1985.

W. Q. Meeker, G. J. Hahn, and L. A. Escobar. *Statistical Intervals: A Guide for Practitioners and Researchers*, Second edition. John Wiley & Sons, Hoboken, New Jersey, 2017.

I. Meilijson. Estimation of the lifetime distribution of the parts from the autopsy statistics of the machine. *Journal of Applied Probability*, 18:829–838, 1981.

R. W. Miller and W. B. Nelson. Optimum simple step-stress plans for accelerated life testing. *IEEE Transactions on Reliability*, 32:59–65, 1983.

D. B. Montgomery. *Design and Analysis of Experiments*, 8th edition. John Wiley & Sons, Hoboken, New Jersey, 2013.

M. D. Morris. A sequential experimental design for estimating a scale parameter from quantal life testing data. *Technometrics*, 29:173–181, 1987.

B. M. Mun, C. Lee, S. G. Jang, B. T. Ryu, and S. J. Bae. A Bayesian approach for predicting functional reliability of one-shot devices. *International Journal of Industrial Engineering*, 26:71–82, 2019.

D. N. P. Murthy, M. Xie, and R. Jiang. *Weibull Models*. John Wiley & Sons, Hoboken, New Jersey, 2003.

H. Nagatsuka and N. Balakrishnan. A consistent method of estimation for the parameters of the three-parameter inverse Gaussian distribution. *Journal of Statistical Computation and Simulation*, 83:1915–1931, 2013.

H. Nagatsuka and N. Balakrishnan. Consistent estimation of parameters and quantiles of the three-parameter gamma distribution based on type-ii right-censored data. *Journal of Statistical Computation and Simulation*, 85: 2406–2424, 2015.

H. Nagatsuka and N. Balakrishnan. Existence, uniqueness and consistency of estimation of life characteristics of three-parameter weibull distribution based on type-ii right censored data. *Journal of Statistical Computation and Simulation*, 86:1248–1279, 2016.

H. Nagatsuka, T. Kamakura, and N. Balakrishnan. A consistent method of estimation for the three-parameter Weibull distribution. *Computational Statistics & Data Analysis*, 58:210–226, 2013.

R. B. Nelsen. *An Introduction to Copulas*. Springer, New York, 2007.

W. B. Nelson. Accelerated life testing - step-stress models and data analyses. *IEEE Transactions on Reliability*, 29:103–108, 1980.

W. B. Nelson. *Applied Life Data Analysis*. John Wiley & Sons, Hoboken, New Jersey, 2003.

W. B. Nelson. *Accelerated Testing: Statistical Models, Test Plans, and Data Analysis*. John Wiley & Sons, Hoboken, New Jersey, 2009.

M. Newby. Monitoring and maintenance of spares and one shot devices. *Reliability Engineering & System Safety*, 93:588–594, 2008.

M. S. Nikulin and R. Tahir. Application of Sedyakin's model and Birnbaum-Saunders family for statistical analysis of redundant systems with one warm stand-by unit. *Journal of Mathematical Sciences*, 188:724–734, 2013.

S. Nowik. Identifiability problems in coherent systems. *Journal of Applied Probability*, 27:862–872, 1990.

C. C. Pan and L. Chu. Reliability assessment for one-shot product with Weibull lifetime components. *International Journal of Quality & Reliability Management*, 27:596–610, 2010.

Z. Pan, N. Balakrishnan, Q. Sun, and J. Zhou. Bivariate degradation analysis of products based on Wiener processes and copulas. *Journal of Statistical Computation and Simulation*, 83:1316–1329, 2013.

L. Pardo. *Statistical Inference based on Divergence Measures*. Chapman and Hall/CRC Press, New York, 2006.

C. Park. Parameter estimation of incomplete data in competing risks using the EM algorithm. *IEEE Transactions on Reliability*, 54:282–290, 2005.

F. Pascual. Accelerated life test planning with independent Weibull competing risks with known shape parameter. *IEEE Transactions on Reliability*, 56: 85–93, 2007.

W. Peng, Y. F. Li, Y. J. Yang, S. P. Zhu, and H. Z. Huang. Bivariate analysis of incomplete degradation observations based on inverse Gaussian processes and copulas. *IEEE Transactions on Reliability*, 65:624–639, 2016.

H. Pham. (*Ed.*) *Springer Handbook of Engineering Statistics*. Springer, New York, 2006.

G. O. Roberts and J. S. Rosenthal. Optimal scaling for various Metropolis-Hastings algorithms. *Statistical Science*, 16:351–367, 2001.

G. O. Roberts, A. Gelman, and W. R. Gilks. Weak convergence and optimal scaling of random walk Metropolis algorithms. *The Annals of Applied Probability*, 7:110–120, 1997.

G. Schwarz. Estimating the dimension of a model. *The Annals of Statistics*, 6: 461–464, 1978.

B. Schweizer. Thirty years of copulas. In G. Dall'Aglio, S. Kotz, and G. Salinetti, editors, *Advances in Probability Distributions with Given Marginals*, pages 13–50. Springer, Dordrecht, The Netherlands, 1991.

N. M. Sedyakin. On one physical principle in reliability theory. *Technical Cybernetics*, 3:80–87, 1966.

M. Shaked and N. D. Singpurwalla. A Bayesian approach for quantile and response probability estimation with applications to reliability. *Annals of the Institute of Statistical Mathematics*, 42:1–19, 1990.

A. Sklar. Fonctions de repartition an dimensions et leurs marges. *Publications de l'Institut de statistique de l'Université de Paris*, 8:229–231, 1959.

A. Sklar. Random variables, joint distribution functions, and copulas. *Kybernetika*, 9:449–460, 1973.

L. J. Slater. *Generalized Hypergeometric Functions, Reissue edition*. Cambridge University Press, Cambridge, England, 2008.

W. B. Thomas and R. E. Betts. Electroexplosive device, 1967.

J. R. Tovar and J. A. Achcar. Dependence between two diagnostic tests with copula function approach: a simulation study. *Communications in Statistics - Simulation and Computation*, 42:454–475, 2013.

S. T. Tseng, N. Balakrishnan, and C. C. Tsai. Optimal step-stress accelerated degradation test plan for gamma degradation processes. *IEEE Transactions on Reliability*, 58:611–618, 2009.

D. Valis, Z. Vintr, and M. Koucky. Contribution to modeling of complex weapon system reliability. In *Proceedings of the European Safety and Reliability Conference*, pages 1813–1818, 2008.

A. Vassilious and A. Mettas. Understanding accelerated life-testing analysis. *Annual Reliability and Maintainability Symposium Tutorial Notes*, pages 1–21, 2001.

R. Viveros and N. Balakrishnan. Statistical inference from start-up demonstration test data. *Journal of Quality Technology*, 25:119–130, 1993.

B. X. Wang, K. Yu, and Z. Sheng. New inference for constant-stress accelerated life tests with Weibull distribution and progressively Type-II censoring. *IEEE Transactions on Reliability*, 63:807–815, 2014.

L. Wang. Inference of constant-stress accelerated life test for a truncated distribution under progressive censoring. *Applied Mathematical Modelling*, 44:743–757, 2017.

W. D. Wang and D. B. Kececioglu. Fitting the Weibull log-linear model to accelerated life-test data. *IEEE Transactions on Reliability*, 49:217–223, 2000.

J. Warwick and M. C. Jones. Choosing a robustness tuning parameter. *Journal of Statistical Computation and Simulation*, 75:581–588, 2005.

H. White. Maximum likelihood estimation of misspecified models. *Econometrica*, 50:1–25, 1982.

C. Xiong, K. Zhu, and M. Ji. Analysis of a simple step-stress life test with a random stress-change time. *IEEE Transactions on Reliability*, 55:67–74, 2006.

W. Y. Yun, Y. J. Han, and H. W. Kim. Simulation-based inspection policies for a one-shot system in storage over a finite time span. *Communications in Statistics - Simulation and Computation*, 14:1979–2003, 2014.

M. Zelen. Factorial experiments in life testing. *Technometrics*, 1:269–288, 1959.

A. Zellner. *An Introduction to Bayesian Inference in Econometrics*. John Wiley & Sons, New York, 1971.

J. R. Zhang, X. Y. Li, T. M. Jiang, and Z. Z. Ge. Optimization of the test stress levels of an ADT. In *Proceedings-Annual Reliability and Maintainability Symposium, RAMS*, pages 1–6, 2011.

X. P. Zhang, J. Z. Shang, X. Chen, C. H. Zhang, and Y. S. Wang. Statistical inference of accelerated life testing with dependent competing failures based on copula theory. *IEEE Transactions on Reliability*, 63:764–780, 2014.

Y. Zhang and W. Q. Meeker. Bayesian life test planning for the Weibull distribution with given shape parameter. *Metrika*, 61:237–249, 2005.

W. Zhao and E. A. Elsayed. A general accelerated life model for step-stress testing. *IIE Transactions*, 37:1059–1069, 2005.

J. Zheng, Y. Li, J. Wang, E. Shiju, and X. Li. Accelerated thermal aging of grease-based magnetorheological fluids and their lifetime prediction. *Materials Research Express*, 5:085702, 2018.

Author Index

Accelerated Life Testing of One-shot Devices: Data Collection and Analysis, First Edition.
Narayanaswamy Balakrishnan, Man Ho Ling, and Hon Yiu So.

Companion Website: www.wiley.com/go/Balakrishnan/Accelerated_Life_Testing

Subject Index

Accelerated Life Testing of One-shot Devices: Data Collection and Analysis, First Edition.
Narayanaswamy Balakrishnan, Man Ho Ling, and Hon Yiu So.

Companion Website: www.wiley.com/go/Balakrishnan/Accelerated_Life_Testing

Printed and bound by CPI Group (UK) Ltd, Croydon, CR0 4YY